UNIT

OCR

AS | F762

Geography

Managing Change in Human Environments

Peter Stiff

With my deep appreciation of Alison's forbearance and encouragement

Philip Allan Updates, an imprint of Hodder Education, an Hachette UK company, Market Place, Deddington, Oxfordshire OX15 0SE

Orders
Bookpoint Ltd, 130 Milton Park, Abingdon, Oxfordshire, OX14 4SB
tel: 01235 827827
fax: 01235 400401
e-mail: education@bookpoint.co.uk
Lines are open 9.00 a.m.–5.00 p.m., Monday to Saturday, with a 24-hour message answering service. You can also order through the Philip Allan Updates website: www.philipallan.co.uk

© Philip Allan Updates 2009

ISBN 978-0-340-94792-0
First printed 2009
Impression number   5   4   3
Year   2014   2013   2012   2011

This guide has been written specifically to support students preparing for the OCR AS Geography Unit F762 examination. The content has been neither approved nor endorsed by OCR and remains the sole responsibility of the author.

Typeset by Pantek Arts Ltd, Maidstone
Printed in India

Hachette UK's policy is to use papers that are natural, renewable and recyclable products and made from wood grown in sustainable forests. The logging and manufacturing processes are expected to conform to the environmental regulations of the country of origin.

AS Geography

Contents

Introduction

■ ■ ■

Content Guidance

■ ■ ■

Questions and Answers

Introduction

About this guide

This guide is designed to help you prepare for OCR AS Geography **Unit F762: Managing Change in Human Environments**.

This **Introduction** explains the assessment structure and outlines the techniques for dealing with structured, extended-writing and essay questions.

The **Content Guidance** section outlines the specification content and the key themes used to formulate examination questions.

The **Question and Answer** section provides eight specimen questions (four structured questions and four extended-writing questions) and two student answers for each question, ranging from grade A to grade E. The student answers are followed by examiner's comments.

Assessment

F762 Managing Change in Human Environments is one of two units that make up the AS specification. It is worth 100 uniform marks, and accounts for 50% of the specification weighting.

The other unit is F761: Managing Physical Environments (see Table 1).

Table 1 AS Geography: scheme of assessment

Unit number and unit name (exam length)	Raw marks	Uniform marks (AS weighting)
F761 Managing Physical Environments (1½ hours)	75	100 (50%)
F762 Managing Change in Human Environments (1½ hours)	75	100 (50%)

Unit F762 covers four major environments:
- Managing urban change
- Managing rural change
- The energy issue
- The growth of tourism

The exam paper will be in two parts — Section A and Section B.

In Section A you have to answer **two** structured questions, You have to answer **one** from **either** Managing urban change **or** Managing rural change **and one** from **either** The energy issue **or** The growth of tourism.

Section B requires you to answer **one** extended-writing question. There will be one question for each of the four environments. The question you answer in Section B must be on a different environment from the two you chose in Section A. The requirement of the question paper means, therefore, that you must study at least three of the four environments specified for Unit F762.

Structured questions

The structured questions in Section A are divided into four sub-questions, worth 4, 6, 6 and 9 marks respectively. Two of these sub-questions require knowledge of one or more geographical examples, and two are usually linked to stimulus materials such as OS maps, charts, photographs and diagrams.

Structured questions are worth 50 out of the 75 marks available for Unit F762. Thus, in a 1.5 hour exam, you should devote approximately 30 minutes to each structured question. You should place particular emphasis on the 9-mark sub-question.

Extended-writing questions

Section B requires you to answer an extended-writing or essay-style question. You should allow approximately 30 minutes for this. The questions demand description, explanation, analysis, application and, most importantly, detailed reference to examples and case studies.

Mark scheme criteria

Examination answers are assessed against three criteria or assessment objectives (AOs). For AS Geography these are as follows:

- **Demonstrate knowledge and understanding** of the specification content, concepts and processes.
- **Analyse, interpret and evaluate** geographical information, issues and viewpoints, and apply them in unfamiliar contexts.
- **Investigate, conclude and communicate**, by selecting and using a range of methods, skills and techniques to investigate questions and issues, reach conclusions and communicate findings.

You should study these criteria carefully because they describe how your examination answers will be judged. The section on examination skills below explains how assessment objectives are used in mark schemes to assess your answers. Table 2 shows the weighting given to each AO.

Table 2 Assessment objective weightings in AS Geography

Unit number and unit name	% of AS			
	AO1	AO2	AO3	Total %
AS Unit F761 Managing Physical Environments	25	10	15	50
AS Unit F762 Managing Change in Human Environments	25	10	15	50

Examination skills

Success in AS Geography requires not only sound knowledge and understanding of the specification content, but also effective exam technique.

To acquire a solid knowledge base you should structure your revision around the key ideas and questions set out in the **Content Guidance** section of this guide. This structure will help focus your learning on the areas most frequently targeted by examiners.

To achieve the higher grades in the more demanding questions you must be able to apply your knowledge and understanding accurately and in unfamiliar contexts. Failure to apply these to a question is a common cause of low marks.

Answering structured questions

Structured questions have a gradient of difficulty. The initial sub-questions are less demanding and carry fewer marks than later questions. The early sub-questions often use command words such as 'describe' or 'outline', while later questions may focus on analysis, explanation and examples.

Stimulus materials are used both directly and indirectly. For direct use, OS maps and photographs are provided to assess key skills, such as map reading and interpretation. Charts and sketch maps may be supplied to assess your ability to summarise, recognise, describe and analyse spatial patterns and trends. For indirect use, stimulus materials are presented as catalysts for assessing wider knowledge and understanding.

Mark schemes for structured questions are levels-based. There are two levels of attainment for 4- and 6-mark questions, and three levels for 9-mark questions. Marks tend to be weighted towards the top end. For example, in a 9-mark question, 8 or 9 marks will be reserved for a level 3 answer. Each level is defined by a descriptor (see Table 3). Having read your answer, the examiner will first allocate it to a level, and then decide the precise mark.

Table 3 Basic descriptors used to assess structured questions

4-mark questions		
Level	Mark	General descriptor
2	3–4	Clear descriptions; specific use of data; accurate terminology
1	0–2	Basic descriptions; generalised; inaccurate terminology
These might be point marked depending on the precise wording of the question		
6-mark questions		
Level	Mark	General descriptor
2	5–6	Clear understanding; detailed explanation; casual links clearly explained; accurate terminology
1	0–4	Limited understanding and explanation; links may be stated rather than explained; inaccurate terminology

9-mark questions		
Level	Mark	General descriptor
3	8–9	A good range of clear, detailed and valid reasons/factors/causes etc; well-chosen example(s); well-structured; accurate grammar and spelling; accurate terminology
2	5–7	A number of valid reasons/factors/causes etc; clearly identified example(s); may have poor structure; some inaccurate grammar and spelling; some inaccurate terminology
1	0–4	Descriptive observations; no links established; limited/no example(s); basic communication; little or no structure; inaccurate terminology

When answering structured questions, you should follow these guidelines:
- Read all parts of the question before attempting to answer. This will help you to avoid repetition in later answers and allow you to get an overview of how the topic should be developed.
- Study the stimulus material carefully.
- Make sure you understand precisely what each question is asking you to do.
- For the 9-mark question, which may require up to 20 lines, (about one side of the answer booklet) you will need to plan your answer. Make a list of the key points and the specific examples you want to include in your answer.
- Divide your time realistically and adjust the length of your answers to the mark weighting. A 4-mark question will require no more than six lines, whereas for a 6-mark question you will need to write approximately eight to nine lines.

Answering extended-writing questions

Answers are assessed against three criteria (or assessment objectives). Each assessment objective is divided into three attainment levels, with a maximum of 13 marks for knowledge and understanding, 5 for analysis and application, and 7 for skills and communication (see Table 4). The relatively large weighting given to skills and communication underlines the importance of accurate spelling and grammar, as well as the ability to structure your answer and provide a clear conclusion.

Table 4 Basic descriptors used to assess extended-writing questions

Knowledge and understanding		
Level	Mark	General descriptor
3	11–13	Detailed knowledge and understanding; cause and effect is well understood and there is effective use of detailed exemplification
2	7–12	Some knowledge and understanding; cause and effect is understood and there is use of exemplification
1	1–6	Limited knowledge and understanding; cause and effect is not well understood and there is limited exemplification. If no located example then top of Level 1 maximum

Analysis and application		
Level	Mark	General descriptor
3	5	Clear interpretation and application of knowledge and understanding to the demands of the question
2	3–4	Some interpretation and application of knowledge and understanding to the demands of the question
1	0–2	Limited interpretation and application of knowledge and understanding to the demands of the question
Skills and communication		
Level	Mark	General descriptor
3	6–7	Answer is well-structured; effective use of grammar and spelling; accurate use of terminology; clear conclusion
2	4–5	Answer may have poor structure; some inaccurate grammar and spelling; inaccurate terminology; limited conclusion
1	0–3	Basic communication with little or no structure; inaccurate spelling; no conclusion

Extended-writing questions can be based on any of the 'questions for investigation'.

Extended-writing questions have a number of common features:
- They require description and explanation.
- They require interpretation and application of knowledge and understanding to the question.
- They always require exemplification using at least two (often contrasting) geographical case studies.

Figure 1 shows a typical extended-writing question and illustrates how the question provides opportunities to assess knowledge and understanding, and analysis and exemplification.

Figure 1 The main features of an extended-writing question

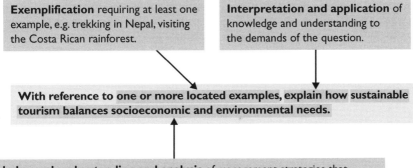

Exemplification requiring at least one example, e.g. trekking in Nepal, visiting the Costa Rican rainforest.

Interpretation and application of knowledge and understanding to the demands of the question.

With reference to one or more located examples, explain how sustainable tourism balances socioeconomic and environmental needs.

Knowledge and understanding and analysis of management strategies that are designed to promote sustainable tourism which takes into account the need for wealth creation, care of local culture and conservation of the physical environment.

You should spend 2 or 3 minutes before you start to answer each question thinking about it and writing a brief plan of your answer. There is space for your plan in the answer booklet. Your plan should outline the general content of each paragraph of the answer and the geographical examples you intend to use to support your answer. An effective plan has three main components: an introduction, a main body and a conclusion. These components will be exemplified here by referring to the question in Figure 1:

- **Introduction** The introduction should define any key terms used in the question, such as 'sustainable' and 'socioeconomic needs', and should indicate the broad structure of your answer. In this example, you might identify some examples of socioeconomic and environmental needs. The introduction should be brief and to the point.
- **Main body** This is where you develop the list of points in your introduction. In one paragraph, for example, you could show how tourism brings much wealth creation to areas and that sustainable tourism tries to minimise leakage. The example of the Annapurna Project in Nepal or some of the eco-lodges in various rainforest locations would be helpful here. In another paragraph, you could explain how some environmentally fragile locations are limiting tourist numbers and opting for a small number of high-value visitors. This strategy aims to protect the physical environment but still maintain the socioeconomic benefit of tourism. Examples from the Himalayas are relevant as are many of the specialist holidays focused on wildlife observation. You should try to point out how the various strategies balance what are often seen as competing needs, for example the operations of various national park authorities such as in USA and England and Wales.
- **Conclusion** This should be a brief summary of the points developed in your answer. In this example, it might be worth concluding that balancing different needs can lead to conflicts but not in all circumstances.

Command words and phrases

Command words and phrases in examination questions are crucial because they tell you exactly what you have to do. You must respond precisely to their instructions. For example, the instruction 'describe' is very different from 'explain'. Ignoring command words and phrases is a common error, and is a major cause of under-achievement. Table 5 lists some typical command words and phrases used in questions in the OCR AS Geography examination and explains what they require you to do.

introduction

Table 5 Key command words and command phrases

Command word/phrase	Requirements
Describe	Provide a word picture of a feature, pattern or process. Descriptions in short-answer questions are likely to be worth 4 or 6 marks and will require some detail.
Outline	The same as 'describe' but requiring less detail. The idea is to identify the basic characteristics of a feature, pattern or process.
Compare	Describe the similarities and differences of at least two features, patterns or processes.
Examine	Describe and comment on a pattern, process or idea. 'Examine' often refers to ideas or arguments that demand close scrutiny from different viewpoints.
Why?/Explain/Account for/ Give reasons	Explain the causes of a feature, process or pattern. This usually requires an understanding of processes. Explanation is a higher-level skill than description and this is reflected in its greater mark weighting in examination questions.

Case studies

An important feature of the OCR AS Geography specification is its emphasis on exemplification through detailed case studies. All the extended-writing questions in Section B, and the 9-mark structured question in Section A, require examples that refer to specific geographical areas.

The extended answer questions often require examples that demonstrate the importance of sustainable approaches to management. Generalised answers to these questions will not achieve the highest marks. Your revision of content for each topic must, therefore, always include at least one, and sometimes two, case studies.

Many case studies are appropriate for more than one topic. For example, a case study of tourism on the Costa del Sol could be used in both the 'Coastal environments' option of Unit F761 and in the 'Growth of tourism' option in this unit; and a case study of the Colorado River could be equally applicable to the 'River environments' option in Unit F761 and to the 'Energy issue' option in this unit.

Content Guidance

This section provides a summary of the key ideas and content detail needed for AS Geography Unit F762 Managing Change in Human Environments.

The content is divided into four main areas:

- Managing urban change
- Managing rural change
- The energy issue
- The growth of tourism

When you revise it is important to use a framework that reflects how examiners might test your knowledge and understanding. Therefore, in addition to the key ideas and content detail, this section provides key questions and answers for each topic.

You should study the questions and answers carefully and organise your revision around them. Focusing on the key questions and adding details of your own to the answers should give you a head start in the final examination.

It is essential that you learn the terminology used in the answers to the key questions, particularly the words in **bold** type. You must be able to apply these terms appropriately in your exam answers.

Managing urban change

What are the characteristics of urban areas?

Key ideas	Content detail
Urban areas have a variety of functions and processes influencing them	The study of two urban areas to illustrate that: • Urban areas are distinct from rural areas in the functions they possess • Urban areas are the product of the interaction of a number of processes
Patterns of land use are influenced by a number of factors and these vary from place to place	Patterns of urban land use in MEDCs and LEDCs

Key ideas

What is meant by urban?

No internationally agreed definition of 'urban', 'town' or 'city' exists. Individual nations have their own definitions; in France 2000+ people are required for a place to be urban; in India it is 5000+. Other characteristics are also used such as percentage of workers in non-agricultural occupations or population density. The lack of a standard definition means that general statements about global patterns must be viewed with caution. Comments about lifestyle, values and attitudes of people can be considered useful when distinguishing between urban and rural.

What functions are present in urban areas?

Functions are activities and a wide range is carried out in most urban areas. Functions serve the urban area's inhabitants and also people living in the surrounding area. Depending on the importance of the city, some functions can have national and international links: finance, fashion and media for example. In broad terms key urban functions are residential, economic (e.g. manufacturing), commercial and retail, political and administrative, other public-service related (e.g. schools, hospitals), involved with transport infrastructure, and recreational.

Different combinations of functions are found in different urban areas. Where a particular function dominates, much of the urban area reflects this. **Land-use patterns** are a visual indication of such specialisation, for example tourism in Bath and Blackpool and manufacturing in Belo Horizonte, Brazil. Employment and way of life in terms of daily, weekly and annual rhythms will also reflect this dominance.

How do economic processes affect urban land use?

Where there is a free market for land, whoever is prepared to offer or bid the most money will usually acquire the land. Some commercial land users are prepared to bid more for a central location than other potential users as it gives them the best access to customers. Traditionally, the city centre has been the most accessible location compared to other urban areas. Transport routes converge on the centre, with the built-up area spreading more or less evenly around it. Once this location operates as the centre, it acquires other advantages such as certain types of building, prestige and specialist services. Residential land use cannot compete with commercial use in central locations and so it bids for land further away. Many cities have grown as housing has out-bid agriculture for land on the outskirts of the urban area.

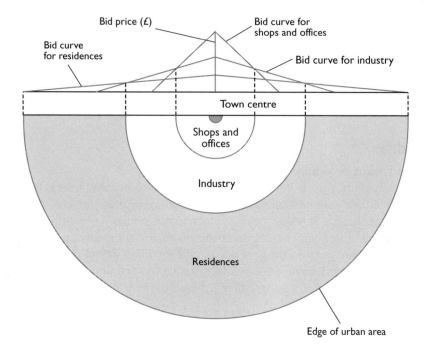

Figure 1 Bid rent model

The **land value surface** can be identified as a three-dimensional aspect of urban areas. It highlights the general decline in land values with distance from the central business district (CBD) but also illustrates subordinate peaks of land value. These tend to be found where access is improved such as at an intersection of main roads in suburbs. A cluster of shops and small offices can indicate such a location.

Among residential areas there are clear differences according to cost of housing. In general there is a decline in housing cost with increasing distance from the centre. This leads to the apparent paradox of cheaper houses in the inner areas being located on more expensive land, and the reverse occurring in the outer suburbs with expensive houses on cheaper land. The solution lies mostly with the density at which the housing is built. The inner areas are built at much higher densities than the suburbs where more houses have land around them in the form of gardens and driveways. There are exceptions, such as high-cost locations near urban centres. Sometimes these are longstanding patterns, as in London where parts of Kensington and Chelsea retain high-status housing as these neighbourhoods benefit from being close to the high-order functions of government and the City of London. Other inner areas have been gentrified, that is, an influx of higher social-status groups has bought run-down properties and renovated them, for example St Peter's Wharf, Sunderland, and La Boca, Buenos Aires. There are also lower value areas on the out-skirts in the form of local authority estates such as Seacroft in Leeds.

How do social processes affect urban land use?

Where people live within an urban area is sometimes seen in terms of the location of different social groups. There are a number of ways of allocating people into social groups, such as income or ethnicity. When these are put together the pattern that results is in the form of a **residential mosaic**. Different social groups tend not to live in the same area; they are segregated from each other across the space of the settlement.

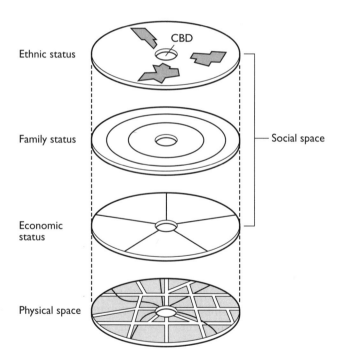

Figure 2 Residential mosaic

Individual decisions that together make up the pattern of spatial segregation are strongly influenced by people's incomes. In the UK nearly two-thirds of all house-holds are **owner-occupiers**, that is, they either own outright or are buying their home. To achieve this most people need to borrow money in the form of a mortgage and their salary largely determines how much a bank or building society is prepared to lend. Lending money to people in some occupations, such as professional and managerial jobs, is considered less risky. Higher incomes also enable people to spend more on mobility, so they can own cars and pay the higher costs of commuting from the suburbs.

Many people move through a **life cycle** that involves changing their type of accommodation and therefore where they might live within an urban settlement. Such moves are often associated with alterations to income levels with age and/or with changes in household size, for example at marriage or with the birth of children. When a young person leaves home to set up their first independent household they usually have limited income and do not need much space so that they often live in a relatively cheap, central flat. A couple with children might try to buy a larger house with more space and a garden in the suburbs, while in retirement the demand for space is reduced and people may move to smaller accommodation or migrate away from the urban centre.

Not everyone follows the cycle and the high rates of divorce in many MEDCs result in a more complex set of housing needs. There are also locational differences throughout the stages depending on whether the household is high or low income.

Within urban settlements many **ethnic groups** are highly segregated. Generally, the more different the group is from the majority population the greater the degree of **segregation**. Where the segregation is very pronounced, ghettoes can develop.

Segregation occurs due to a number of factors, some of which are positive while others are negative. Positive reasons include people's desire to share their home language, to share their own places of worship and schools, and to have access to specialist shops such as food suppliers. Negative reasons include defence and security against prejudice from the host society, a low economic status allowing little choice in the housing market that forces people into low-cost areas such as inner-city terraced housing, and discrimination in the housing market.

Over time some ethnic groups are assimilated into the host society and so the individual households tend to disperse away from the original cluster.

How do political processes affect urban land use?
In some circumstances land does not have a monetary value and cannot be bought or sold. Many east European countries followed a communist/**centrally-planned** system for much of the second half of the twentieth century under which land was owned by the state. Urban developments were not subject to market forces and other considerations. The effect of this could be seen in land use such as the construction of large-scale public housing projects and areas of heavy industry. Countries such as North Korea continue to be centrally planned.

Many MEDCs have a mixed economy of private and state influence. Under such a system **public housing** can play a significant part in urban development. Most UK cities have extensive areas of local authority housing on their outskirts or in areas of comprehensive redevelopment in the inner city. In the 1980s and 1990s much publicly owned housing was sold off and so the amount of housing available for rent from local authorities fell. Some other land uses are also publicly owned such as hospitals, schools and parks. The planning process also represents a political force, for instance when a green belt is placed around a settlement.

In some LEDCs the clearance of slum areas has political origins. In Nairobi, past governments have refused to recognise **informal settlements** that were considered illegal. Neighbourhoods were demolished and their inhabitants forced to move elsewhere. More common now is a political acceptance of slum areas with various improvement schemes being implemented.

What influences can physical factors have on urban land-use patterns?

When an area is covered by buildings, roads and railways, it can be difficult to see the influence of the physical landscape. Ordnance Survey maps at 1:50 000 and 1:25 000 scales can reveal some physical factors as can satellite images. Computer-generated maps of potential flood risk are useful sources of information showing the influence of this physical factor on land use.

If an urban area is located on the coast, such as Sunderland, then it can only develop inland from and along the coastline. This will modify the arrangement of land use so that if a series of concentric zones develop, they will exist as half circles. The presence of a river can encourage sectors to develop along it. Floodplains tend to repel development, although with increased flood prevention in MEDCs, some land uses such as industry and warehousing have been attracted to the cheaper land in such locations. In MEDCs steep slopes also repel developments, although hills within urban areas often attract higher-status housing. There is increased competition for the sites that offer good views and a chance to be above the pollution of the rest of the city.

In many UK towns and cities there is a marked west–east contrast in housing areas, with locations to the west being more favoured and therefore more expensive than the east. The westerly prevailing winds blow the pollution generated by the urban area from west to east.

In LEDCs the poor often have little choice but to occupy land that is marginal. Examples of such areas are tidal swamps, floodplains and steep slopes. The slum area of Kibera, Nairobi, lies mostly on the floodplain of the highly polluted Ngong River.

How is land use distributed in MEDC urban areas?

Because the processes operating within urban areas are not random, neither are the spatial patterns of land use within towns and cities. Although each urban place has distinctive features, it is possible to recognise basic patterns. These **models** are important in identifying and understanding fundamental processes. Without this framework, sustainable management can become ineffective.

Many MEDC cities are focused on a central business district (CBD). The land use around this can be arranged as a series of concentric rings or **zones**. Cities that grew rapidly from a central point tend to have each new phase of growth added on the outside of the previous ring. The process of **invasion and succession** helps explain the pattern. Each new group of migrants settled first in cheaper housing near the CBD. As they became economically and socially more secure they moved out into the next ring. The group already there continued their upward economic and social mobility and so moved out themselves.

An alternative arrangement shows land use radiating from the city centre as a series of **sectors**. These sectors tend to develop along a linear feature such as a river or a man-made transport route, for example a main road or railway. There is, therefore, a strong directional element in this arrangement. Within an individual sector the older land use tends to be towards the centre with more recent developments of the same type of land use built towards to outskirts. The process of **filtering** helps explain this pattern. New housing is built for high-income groups. They leave their old houses which are then occupied by middle-income groups. Over time housing filters down the economic and social hierarchy.

By the later twentieth century, most MEDC cities had several foci where economic activity was concentrated. Such **multiple nuclei** can include parts of the city specialising in industrial, retail or office activities. This model assumes a high degree of personal mobility as people have to travel among the respective nuclei. It also highlights that some land uses repel each other, for example polluting industries such as petrochemicals and high-status housing, while others attract, for instance open recreational space and high-status housing.

Within many UK urban places a combination of patterns exists. This reflects processes that were operating in the past and have left a physical legacy which present-day planning and management must take account of.

How is land use distributed in LEDC urban areas?

There is much diversity among these cities due to their contrasting historical development, including their different experiences of colonialism, and their present-day economy and culture. However, some common features can be recognised. A commercial core area is quite common as many of the larger cities were developed on the basis of trade; often this is associated with port functions. If there had been a strong colonial influence, walled forts and clearly defined areas of low-density housing for the administrators can be identified. This housing was often taken over by indigenous government officials after independence. Traditional commercial centres, such as bazaars, are common in areas of unplanned higher-density housing, usually located away from the higher-status areas.

More recent industrial developments are usually found at the edge of the built-up area or close to the port where squatter settlements develop to house those working in the factories.

What are the social and economic issues associated with urban change?

Key ideas	Content detail
Urban growth and decay can lead to a variety of social and economic issues in urban areas	The study of two contrasting urban areas (e.g. Harehills and Seacroft in Leeds) to illustrate: • The social and economic differences existing in urban areas • That a significant proportion of urban dwellers suffer social and economic deprivation • The characteristics of urban deprivation include low levels of economic well-being, housing and environmental quality and social conditions • The problems of managing the growing demand for services such as health, education and public transport

Key ideas

How and why do urban areas experience growth or decay?

Urbanisation is the process whereby the proportion of people living in towns and cities increases. When a society is undergoing urbanisation there is a relative shift of population from rural to urban places. **Urban growth** is the increase in the number of people living in urban areas or an increase in area of towns and cities.

Both these processes were very active during the late eighteenth, nineteenth and early twentieth centuries in MEDCs. They were associated with changes from agrarian societies to those dominated by manufacturing and the development of tertiary activities.

In the latter twentieth century many of the larger urban areas in MEDCs experienced **counter-urbanisation**. Population numbers fell as people moved out of large cities into smaller towns and villages within commuting distance of the city. There was a degree of employment migration and the setting up of new economic activities in the non-metropolitan locations. Inner-city areas lost much of their economic activity with some of the loss due to **economic globalisation**. Planning and management schemes were implemented to clear slum areas, replacing high-density housing with lower-density styles.

More recently something of an **urban renaissance** can be seen in many cities. Waterfront locations and former industrial and warehousing sites have been redeveloped as housing locations. The River Thames through much of London is now lined with apartments, and smaller cities such as Sunderland have seen similar developments, for example at St Peter's Wharf. This **post-industrial** phase is characterised by a shift in emphasis of urban areas from secondary to tertiary activities.

In LEDCs urbanisation and urban growth have proceeded rapidly from the late twentieth century onwards. Between 1982 and 2002, Kenya's urban population increased at an average rate of 6.3% per year; Kenya's population growth rate for the same period was around 2.6% per annum. Both **rural–urban migration** and **natural increase** are responsible for such high rates of urban growth. Nearly twenty cities in the world have over 10 million inhabitants, the majority of these places being located in LEDCs. Economic globalisation has led LEDCs and NICs to experience substantial growth in manufacturing, most of which is carried out in urban locations.

What are the social and economic issues resulting from growth or decay?

Some of the issues are essentially the same whether growth or decay are being experienced. Housing is one example. When urban areas grow rapidly, as in LEDCs today, insufficient housing is available; when urban areas decay, as in some MEDCs today, housing quality can be a serious concern. Likewise, employment becomes an issue when population growth outstrips job increases, while decay leads to high and persistent levels of unemployment.

Health care and education can be both adversely and positively affected by growth and decay, depending on the ability of urban authorities to provide these services. This is often a question of whether sufficient resources are available to allow supply to meet demand.

Poverty and prosperity are usually found when growth or decay is occurring.

What is meant by socioeconomic deprivation?

Deprivation is, quite literally, the absence of something. **Socioeconomic deprivation** refers to people who are disadvantaged as they do not have certain elements in their lives. In the UK an index of **multiple deprivation** combines seven components which indicate the level of poverty: income, unemployment, health/disability, education, skills/training, housing quality, living environment/crime data. The components are interlinked, with change in one feeding back change to other factors.

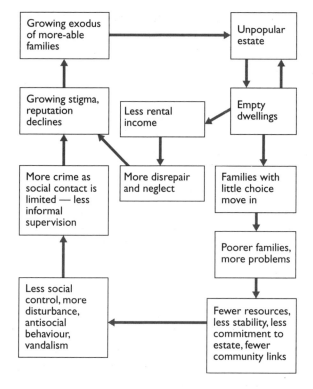

Figure 3 Some of the linkages contributing to multiple deprivation

The primary cause of deprivation is economic but social issues go hand in hand with the economic factors. Low wages are often found in employment sectors where periods of unemployment and/or part-time working are commonplace.

In LEDCs many people are **under-employed**; for example they may work long hours selling items at the roadside, but are not economically productive as they sell only a few per day. With a lack of income comes poor diet, which can lead to an increased susceptibility to ill-health. Ill-health can mean frequent days off work with a consequent reduction in income. Low income causes people to live in inadequate housing, whether in MEDCs or LEDCs, for example Harehills, Leeds, and Kibera, Nairobi.

Lack of money can often lead to poor education. Children leave school to begin earning money for the family; where parents have limited formal education they may not give high priority to children obtaining skills.

Describe and explain the management problems in providing services such as health, education and public transport

Both growth and decay in urban areas can cause difficulties in the provision of services. Authorities in cities that are growing rapidly tend to have difficulty keeping up with rising demand for services. In cities experiencing decay managers can also struggle to maintain a good level of services: as wealth creation reduces, income declines but demand from poorer groups rises. Where services are needed is also an issue. No matter what the situation, poorer urban dwellers have lower levels of **personal mobility**. This has a profound influence on their ability to access services and employment.

In many LEDCs, demand far exceeds the capacity of public and private providers. Slum areas have grown haphazardly over the years making it very difficult to bring in efficiently functioning services such as clean drinking water and sewage systems. Governments lack the resources to train and pay for personnel to work in sectors such as health and education. The capital investment required for public transport schemes such as mass transit programmes, or simply paving roads, is more than governments can afford. Too often schemes are begun but not completed. There are, however, examples of improvements in conditions for urban dwellers, and not just for the wealthy. Many of these directly involve **non-governmental organisations** (NGOs) such as Christian Aid, Oxfam and Save the Children.

The past 50 years have witnessed many schemes in MEDCs seeking to improve people's access to services. To begin with there tended to be **top-down** approaches. Local authorities decided what services were needed and where. Sometimes these succeeded but not in every case. More recently the management strategy has been **bottom-up**. Consultations with local residents, employers, developers, local councils and agencies such as primary health care trusts have resulted in regeneration projects, as in east Leeds. The intention is to create a more holistic and sustainable programme.

What are the environmental issues associated with urban change?

Key ideas	Content detail
Urban change can put increasing pressures on the environment including: • Traffic congestion • Atmospheric pollution • Water pollution • Urban dereliction • Waste disposal	The study of two contrasting urban areas (e.g. London, UK, and Santiago, Chile) to illustrate: • Urban change brings environmental pressures • These pressures vary from one city to another • Increasing resources are invested in managing the environmental pressures

content guidance

Key ideas

What factors contribute to environmental pressures in urban areas?

As urban areas change, pressure is placed on the environment. MEDCs tend to have planning and management systems in place that supply safe water and deal effectively with sanitation and drainage. Air pollution, however, continues to be a concern, though its nature has changed. Changes in economic systems resulting in loss of manufacturing have led to issues about what has been left behind on abandoned factory sites. Of growing anxiety is the issue of waste disposal of all types.

In LEDCS unplanned and spontaneous growth has resulted in urban systems which struggle to cope with most environmental issues. Water, air and land pollution is common. The globalisation of economic systems has also meant that much manufacturing previously located in MEDCs is now in LEDCs or NICs. The goods produced are sold in MEDCs and it is argued that this arrangement simply relocates MEDC pollution to other parts of the world.

What are the issues of managing problems of traffic congestion and air pollution?

The nineteenth and first half of the twentieth centuries saw the high levels of air pollution in MEDC cities. Widespread use of coal for heating and power generation in both domestic and industrial premises led to serious **episodes of air pollution**. The release of **gases** such as sulphur dioxide, carbon monoxide and nitrous oxides was accompanied by high levels of **suspended particles**. Most MEDCs have managed this element of their air pollution effectively so that **smogs** are much less frequent than they used to be. The requirement to use smokeless fuels, the closure of inner-city factories and the change in types of manufacturing have made major contributions.

However, where local circumstances, physical and human, combine in certain ways, air pollution levels can still exceed safe limits. High levels of motor vehicle use, burning of fossil fuels in factories and power plants, and local relief and atmospheric conditions can trap pollutants. Particularly aggravating physical factors include an urban area that is sited in a valley surrounded by significant hills or mountains, where **anticyclonic conditions** contribute to atmospheric inversions and generally light winds, thus hindering dispersal of pollutants. Santiago, Los Angeles and Rome are examples of urban areas regularly suffering from high levels of air pollution due to such interactions.

Most LEDCs continue to struggle in their management of air pollution. Much of the urban population relies on fuels for heating, cooking and hot water that produce high levels of air pollution such as charcoal or paraffin. It is difficult to regulate industrial pollution, especially amongst the many small-scale businesses. Vehicles in LEDCs are often poorly maintained, leading to high levels of polluting exhaust emissions. The growth in disposable incomes among some sectors of LEDC/NIC societies is resulting in rising car ownership levels. This further threatens attempts to reduce air pollution as well as increasing traffic congestion.

It is the continuing impact on health, especially of the very young and elderly, and the production of acid rain that make urban air pollution a serious issue.

The issue of **traffic congestion** affects all urban areas. It reflects the interaction of a combination of factors, most of which are common to many towns and cities. Urban areas often rely on transport infrastructure (roads and railways for example) that was planned and built under different socioeconomic conditions from those of today. The expansion of urban areas in the nineteenth and early twentieth centuries took place when canals, steam railways and horse-drawn vehicles dominated and when most people walked to work, school, the shops and recreation. Thousands of houses were built without space for off-street vehicle storage and along streets wide enough only for the traffic of the day. Some cities retain **mass transit** rail systems from the late nineteenth century, London, Paris and New York for example. These continue to play an important role in managing intra-urban movement and have seen significant recent investment.

The growth in **private-car ownership** poses management problems of storage at origins and destinations as well as of actual movement. Road networks within urban areas are notoriously difficult to manage. Building new and/or improved roads has to take into consideration existing elements of the built environment; demolition of buildings such as housing, schools or hospitals is neither possible nor desirable. Most cities are looking at a combination of controls, road pricing and bus lanes for example, and inducements, such as integration of buses and metros and expansion of mass transit systems. Some cities have had degrees of success such as Santiago with its 10-year Urban Transport Plan.

Increasingly, urban authorities are linking attempts to resolve issues of air pollution and traffic congestion with concerns over greenhouse gas emissions.

What are the issues of managing problems of water pollution?

With socioeconomic development has come increasing demand for water. Some of the earliest examples of metropolitan management of environmental issues were based on providing clean drinking water and safe disposal of contaminated water. Those cities that benefited from nineteenth-century improvements, for instance many in MEDCs, now face issues of renewal of the water infrastructure such as water mains and sewage pipes. Some LEDC cities possess systems built in the late nineteenth and early twentieth century when the countries were administered by colonial powers: these too are in need of renewal, the cities being overwhelmed by population growth, such as Cairo and Kolkata.

Rivers and the sea have always been easy ways to dispose of waste as they tend to move the material to another location — out of sight, out of mind! Most MEDCs today have strict regulations in force to restrict or prevent water being abused in this way. Sewage is treated and industries are made to clean effluent before it can be released; domestic waste water must be discharged into sewers or cess pits, not into open water courses. When and where accidents occur or the law is broken, river authorities tend to be quick to act to deal with pollution episodes. Physical removal of waste and aeration are frequently employed. There are, however, local concerns regarding pressure on neighbourhood hydrological systems with the intensification of land use. For example the paving over of front gardens and increasing building densities in many suburban locations have led to increased local flood risk and the resultant spread of pollution.

LEDCs do not have the funds to manage water resources so effectively. Haphazard disposal of both domestic and industrial waste can lead to water pollution. It is not just the appearance and smell of such pollution; it is the water-borne chemicals and diseases which pose serious threats. **Heavy metals** and **bacteria** and **viruses** are commonly found in many urban rivers and coastal zones. As with other issues, attempts are being made to deal with water pollution but the managers face an uphill struggle. Most large urban areas have projects being either implemented or planned. Municipal authorities in Santiago have ambitious plans to clean up the Mapocho River.

What are the issues of managing urban dereliction?
Because urban areas are dynamic, change is a constant feature. Change in land use can result in a building being abandoned or, after demolition, a plot of land being left unused, both examples of dereliction. The movement of urban functions from one part of the city to another or their migration away from the urban area alto-gether can result in such changes. In addition, some buildings physically decline with increasing age, in particular if they were not well built in the first place. The loss of manufacturing and population from inner areas due to deindustrialisation in the 1970s and 1980s and the decline in much nineteenth-century housing represent changes that can lead to dereliction. The lower Lea valley in east London, Harehills in Leeds and Sunderland's waterfront are examples of locations where urban change has created dereliction.

In most MEDC urban areas **brownfield** sites can be found. These represent dereliction and have been difficult to manage in terms of promoting their re-use. Three factors explain this. Because they have been developed, brownfield sites can be contaminated with pollutants. Topsoil containing heavy metals such as mercury or cadmium from former industrial uses can be particularly difficult (and so expensive) to clean. Second, previously developed sites are often criss-crossed by utilities such as power, water and sewage. Where these are underground their precise routes are not often known and so development costs more. Other services such as roads and rail lines can restrict rede-velopment, for example by restricting access to a site. The third factor is that of the fragmentation of available sites. Large plots are rare, a disincentive for many develop-ers as they will find it difficult to achieve economies of scale.

What are the issues of managing waste disposal?
Both MEDC and LEDC urban areas face problems in managing waste. Generally, more affluent cities generate vast quantities of household waste such as packaging and food waste. Construction/demolition contributes much waste from building sites while offices and factories also have large amounts of waste for disposal. Traditionally, land-fill absorbed waste but sites are rapidly coming close to capacity and new sites are difficult to find. Concerns exist over possible pollution from contamination of

groundwater and methane production from the anaerobic decomposition of organic waste such as food scraps and paper. Some cities export their waste, London for example, but this is expensive. It extends a city's **ecological footprint** and damages environments elsewhere. Incinerators burn waste and can generate electricity and hot water for heating. Concerns about air pollution, visual impact of the plant and impacts of lorry traffic to and from the incinerator mean incinerators are not always viewed positively.

Increasingly **recycling**, including composting, is seen as vital. Interestingly, many LEDCs already recycle much of their urban waste. Large numbers of people make their living from collecting and re-using waste items although there are significant health risks associated with this employment.

How can urban areas be managed to ensure sustainability?

Key ideas	Content detail
• Successful sustainable management requires an understanding of the dynamic nature of social, economic and political processes • Sustainable development of urban areas requires a careful balance of socioeconomic and environmental planning	The study of at least one example to illustrate how planning and management practices enable urban areas to become increasingly sustainable

Key ideas

With reference to at least one example, show how the sustainable development of urban areas requires detailed planning and management of socioeconomic and environmental factors

The concept of the sustainable city is beginning to be taken seriously although its achievement remains some way off at present.

China's considerable economic and urban growth over the past 30 years has been at much environmental cost. Urban managers in China are paying increasing attention to the development of a sustainable approach to urban change. On Chongming island in the delta of the Yangtze River close to Shanghai, an eco-city, Dongtan is being developed. A phased plan of growth from 25 000 residents in 2010 to 500 000 by 2030 is designed to be self-sufficient in energy and water, from renewable energy and rainwater purification. A new urban lifestyle will help reduce the city's ecological footprint to one of the lowest urban figures in the world.

In existing urban places, obtaining widespread sustainability is difficult. Many urban areas possess small-scale projects such as the carbon-neutral housing development of Bedzed, southeast London; larger projects are problematic. The lower Lea valley, east London, embodies many of the management issues affecting urban locations. It is part of the Thames Gateway region, for which ambitious regeneration plans exist. A key element in the plans is the 2012 Olympic Games development. This keynote large-scale project has sustainability at the heart of all its planning and management.

Managing rural change

What are the characteristics of rural areas?

Key ideas	Content detail
• Rural areas have a variety of functions and processes influencing them • The range of functions and opportunities are influenced by a number of factors and these vary from place to place	The study of two rural areas to illustrate that: • Rural areas are distinct from urban areas in the functions they possess • Rural areas are the product of the interaction of a number of processes

Key ideas

What is meant by rural?

Areas dominated by **extensive** land uses such as agriculture or forestry are rural. Also included are areas best described as wilderness such as many mountainous regions, northern Scandinavia for example. There is no internationally agreed definition of 'rural' but neither is there one for 'urban'. Different countries use a variety of indicators to define urban and by default non-urban areas. Indian rural places can have up to 5000 people with not less than 25% of the males working in agriculture and with an average population density of not more than 400 persons per square kilometre. In Denmark, 'rural' would describe a settlement of less than 2000 inhabitants.

What functions are present in rural areas?

Functions are activities and in rural areas they vary with the resources found in the region. Some functions serve the inhabitants of the area such as low-order retailing but others are also used by people from urban settlements such as recreation, walking or fishing.

Production of **primary products** from agriculture, forestry, mining and quarrying is common in rural areas. Some of these are often processed in rural areas, dairies and saw-mills for example. A growing function is energy generation such as wind farms. Tertiary functions exist in the form of low-order shops and offices in small settlements. Transport infrastructure runs through rural areas. Tourism and recreation have become increasingly important in many rural regions in MEDCs. Rural areas also function as residential areas with people living in isolated farms, hamlets, villages and small market towns.

How do economic processes affect the development of opportunities in rural areas?

Traditionally, primary activities provide most of the economic opportunities. However, in many MEDC rural regions, agriculture is no longer the main economic process, although it dominates the land use, with rural dwellers working in service activities. Where the rural region is remote, for example northern Sweden, traditional industries can continue to be significant, forestry for example. Increased demand for tourism and recreation has given many rural areas opportunities for wealth creation.

Most LEDC rural regions continue to be dominated by primary economic processes. A wide variety of agriculture — subsistence, commercial, arable and pastoral — takes place. Exploitation of forests and minerals can both create and destroy opportunities in rural areas such as tropical rainforests.

How do social processes affect the development of opportunities in rural areas?

Rural areas are socially and ethnically relatively **homogenous**. Family ties can be stronger and may be restricted spatially. Communities such as villages can often be more cohesive with most people knowing and interacting with one another. Generally rural society is more **conservative** in its attitudes. These characteristics can result in less opportunity for social mobility and is often seen as a motivating force for young adults to migrate away.

A strong trend in many rural areas in MEDCs is the in-migration of retired people. This can reinforce community life as the elderly have time for local activities but it also poses challenges, in particular regarding the support and care of the very elderly.

Rural areas close to large urban centres tend to be affected by **counter-urbanisation**. Urban dwellers move to villages but continue to work and often shop in the city. For some rural areas the increase in the proportion of second homes can have significant social effects, such as effectively depopulating a village for most of the year.

LEDC rural areas are often strongly conservative. Local traditions seek to maintain the status quo and some see this as an obstacle to development.

How do political processes affect the development of opportunities in rural areas?

Agriculture has been directly affected by political processes in many countries. All sorts of political programmes have been initiated that influence types of farming, most notably the **Common Agricultural Policy** (CAP) of the **EU**. Grants and quotas have influenced how farmers have operated their farms. Policies have been aimed at both reducing and increasing production of different products.

Some rural local authorities have implemented programmes involving **key settlements**. These larger villages become the focus of services such as education and health but there is a consequential loss of services in surrounding small settlements.

Regional financial assistance is offered to some regions in order to boost the opportunities available. Västernorrland in northern Sweden receives money from the European Regional Development Fund, some of which has been used to improve transport infrastructure.

What influences can environmental factors have on the development of opportunities in rural areas?

These factors can advance or hold back development in rural areas. High-quality physical environments provide opportunities for recreation and tourism. **National parks** such as the Yorkshire Dales (England), Yosemite (California, USA) and the Masai Mara (Kenya) attract visitors who then demand goods and services thereby providing employment opportunities.

Where the rural environment is protected, development can be restricted. Although many national park organisations aim to attract visitors, they also have responsibilities towards conservation of habitats which leads to the strict control of building and access. Increasingly, sustainable management is seeking to accommodate both protection and opportunity.

What are the social and economic issues associated with rural change?

Key ideas	Content detail
• Structural change can lead to economic and social differences within and between rural areas • Lack of economic opportunities can lead to depopulation and decline	The study of two contrasting rural areas (e.g. Nebraska, USA, and Wensleydale, UK) to illustrate that: • Growth or decline can affect rural areas • Economic and social problems occur in growing and developing rural areas • Economic and social problems occur in declining rural areas

Key ideas

What factors contribute to the economic and social problems occurring in growing and developing rural areas?

Both urban–rural migration and natural increase have led to counter-urbanisation, the increase in the proportion of people living in areas defined as rural. Rural areas and settlements close to major urban centres in many MEDCs have tended to experience significant growth over the past 40 years. Even remote areas have witnessed growth. Orkney, Scotland, and Idaho, USA, have seen population revivals as people trade off higher incomes and easy access to services for a preferred lifestyle and environmental quality. Such demographic turnaround has not occurred in LEDCs. Although some of the original rural migrants do return from the city, large flows of people towards urban areas continues.

Migration to rural areas is caused by a desire for rural living, perceived improved quality of life, lower house prices, opportunities to work from home and better services such as state schools. Improvements in levels of personal mobility through higher levels of car ownership and improvements in transport infrastructure are important.

Developments in electronic communication allow working from home and internet retailing and banking. **Commuter** or **suburbanised villages** develop within commuting distance of the city. However, as demand for property increases, so do house prices, often pricing local young people out. Traffic levels increase along rural roads resulting in congestion and increasing journey times. Local shops and services may not be well supported by in-migrants leading to a loss of local functions.

More distant rural regions can receive **retirement migration**, especially areas with high amenity value such as Devon and Cornwall, UK, and Provence, France. An ageing population lowers the demand for some services, schools for example, leading to their closure. On the other hand, demand for health care increases, putting a strain on the providers.

What factors contribute to the economic and social problems in declining rural areas?

Rural depopulation results from demographic and economic factors. Both **natural decrease** and **net migrational loss** affect many rural areas in MEDCs, especially remote ones. Young people leave rural areas in search of employment, affordable housing and improved social facilities such as entertainment. Median incomes in rural areas are usually much lower than in metropolitan locations. Employment opportunities are often limited with small farms declining in number and mechanisation increasing in large scale agri-businesses.

There have been reductions in the numbers of businesses linked with farming such as seed and feed merchants and machinery suppliers. Declining populations reduce the **thresholds** of shops and services, schools and health care for example. Their closure further reduces employment and makes rural areas even less attractive.

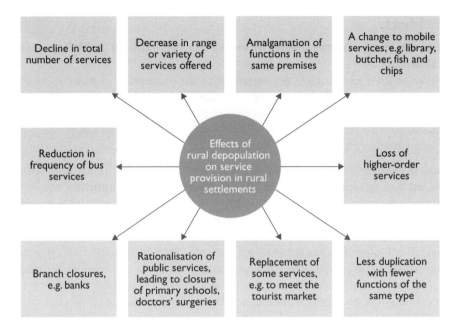

Figure 4 Economic and social problems arising from rural depopulation

In LEDCs employment opportunities may be limited while systems of land inheritance may lead to landlessness. In many rural regions the provision of services such as basic schooling and health care continues to pose significant problems. The impact of **environmental hazards**, both short and long term, can lead to population loss.

What are the environmental issues associated with rural change?

Key ideas	Content detail
The changing use of rural areas puts increasing pressures on the environment including: • Land-use change • Traffic congestion and pollution • Land degradation • Water pollution • Rural dereliction	The study of two contrasting rural areas to illustrate the problems and management of: • Increasing use of rural areas for recreation • Impacts of farming change on the environment • Environmental issues associated with building developments in rural areas • Traffic problems in rural areas

Key ideas

What are the problems caused by increasing recreational and leisure use of rural areas?

Rural recreation and tourism are important to the economies of many rural areas. The visitors, however, bring problems associated with their access to rural areas and with their impact on environments in the course of their recreation and leisure pursuits. Many visitors come by car so that road congestion and car parking have become issues to manage.

In using rural resources, visitors can have serious physical impacts. **Footpath erosion** through trampling can be severe. **Disturbance to habitats** can impact negatively on both flora and fauna. Picking of rare wild flowers such as some orchid species and the interruption of breeding cycles of birds can be significant to survival. In parts of east Africa, hot air balloon flights and vehicle-based safaris have been charged with altering animal behaviour patterns. Where recreation and tourism is water based, bank erosion from speeding boats, and noise and wash disturbance of wildlife can occur. Water quality can be negatively affected by turbulence from power boats. The Broads in Norfolk and Suffolk is one location at risk.

How have these been managed?

The establishment of national parks in both MEDCs and LEDCs is one attempt to manage effectively rural areas of high environmental value. **Sites of Special Scientific Interest** (SSSI), **Areas of Outstanding Natural Beauty** (AONB) and **Ramsar** sites offer further measures of differing protection in the UK. There are hundreds of local nature reserves all trying to conserve ecosystems. Some places restrict access to visitors and most use education, via centres and noticeboards, to encourage people to act responsibly.

Describe and explain the causes of farming change

Political and economic factors are important. Government, both national and supranational, has had a profound effect on agriculture. The EU's CAP used financial means to influence farmers' choices. Guaranteed payments for arable crops such as cereals, sugar beet and oilseed rape meant that these crops dominated. In some regions much pasture land was ploughed and a general **intensification** process took place. Greater use was made of agrochemicals (fertilisers, pesticides and herbicides) and machinery, and the cultivated area increased. **Farm amalgamation** has taken place as smaller farms were absorbed by large-scale agri-businesses. In pastoral regions, milk quotas made it difficult for some dairy farmers to continue in business. Extensive livestock farmers, for example hill sheep farmers, struggled to make their enterprise pay as attention was focused on the arable sector. The situation is, however, dynamic. Adjustments have been made in the system of payments and food prices have been rising. Farming change looks likely to continue.

What have been the impacts of farming change on rural environments?

As economic pressures on farmers mounted, agriculture became more intensive. Increasing specialisation and **economies of scale** have meant that farmers try to use their physical resources to their maximum potential. Ponds and boggy areas have been drained, hedgerows removed and woodland cleared. Where cereal cultivation dominates, very large fields have been created to accommodate the large-scale machinery such farming uses. A prairie-like landscape exists in some parts of regions such as East Anglia. The loss of habitats, such as ponds and hedgerows, results in a decline in **biodiversity**. Shelter, food, breeding sites and corridors for movement are removed. Particularly in lowland areas, habitats such as heaths and wetlands have been under sustained pressure from agriculture. Areas where more land is used for arable now than in the past have witnessed increased sediment runoff from bare fields. Silt ends up in water courses increasing flood risk and lowering water quality. Bare soil can also dry out in winter and become prone to wind erosion.

Pollution of water courses and groundwater has links with changes in farming practices. The increased use of agro-chemicals can lead to nitrates and phosphates leaching into water. The changes water chemistry are toxic to some aquatic plants. The accumulation of nutrients leads to **eutrophication**. The result is often algal blooms and de-oxygenation of water through consequent bacterial activity.

How have these impacts been managed?

Government policies have increasingly taken environmental impacts into consideration. The reform of the CAP, begun in 2000, has sustainable development of rural areas as one its two central 'pillars'. Subsidies and guaranteed prices have been replaced with the **Single Payment Scheme**. This reduces the incentives to farm intensively. In some upland areas, stocking levels have begun to fall resulting in high pastures and moors showing signs of environmental recovery. The **Environmental Stewardship Scheme** was introduced in 2005. Farmers receive payments, the level of which is determined by the environmental sustainability of their management practices. Hedgerows are being planted, stone walls rebuilt and ancient woodlands and wetlands conserved by some farmers. **Set-aside** is a controversial scheme whereby farmers are paid for leaving a proportion of their land fallow. Environmental benefits include the provision of habitats for plants, invertebrates, birds and animals. There are also grants paid to farmers converting land to woodland.

Management of water pollution has focused on raising the standards of water quality discharged from sewage systems and reducing the level of fertiliser runoff. **Nitrate Vulnerable Zones** cover just over half of England, with farmers compelled to take action to limit fertiliser applications.

Describe and explain the causes of building developments in rural areas

Changes in demography and rising standards of living, including incomes, have put great pressure on housing supply. **Increases in life expectancy** and numbers of **single-person households** have added to the demand for housing. Those rural areas within commuting distance of urban centres have experienced many house-building schemes, both small and large scale. Increasing levels of personal mobility, car ownership and road improvements allow people to commute longer linear distances as time distances have been reduced. Planners have been under pressure to allow housing developments to meet the increased demand. The matter of affordability of housing for local inhabitants compared with in-migrants is a concern.

What are the environmental issues associated with building developments in rural areas?

Some developments have been carried out on **greenfield** sites. In some areas, **green-belt** land around urban centres has been developed or is at risk from building. The loss of these habitats reduces biodiversity and fragments rural ecosystems.

The conversion of former agricultural buildings and/or local shops to residences also changes the social environment. The character of villages alters as fewer people have links with agriculture and they may not support community life. In some locations, redundant brownfield sites such as derelict agricultural-based factories have been redeveloped enhancing the environment.

New housing estates may result in increased surface runoff following heavy rain. This raises the flood risk along local streams and rivers.

How have these been managed?

Different levels of government, as well as voluntary organisations, attempt to manage the issues. Planners establish priorities for development and work with local people to try to bring about appropriate decisions. There is also pressure from regional and central government to increase the supply of housing, in particular social housing aimed at those who need to rent.

Describe and explain the causes of traffic problems in rural areas

Main routes cross rural areas carrying through traffic. These routes have a strategic national importance but also have regional and even local significance. With the growth in road traffic many rural settlements experience major disturbance from vehicle flow. Noise, air pollution, vibration damage to buildings and physical danger to pedestrians can be very serious. Counter-urbanisation has increased commuter traffic using rural roads with rush 'hour' twice a day causing much congestion.

The increasing scale of agricultural vehicles can lead to difficulties when farmers need to move between fields.

How have these been managed?

Motorway construction, the upgrading of roads to dual carriageways and construction of by-passes are frequently employed to solve traffic management issues. Where these are not funded, measures aimed at improving traffic flows, and the safety of motorists and non-road users are implemented. Lower speed limits, re-aligned junctions and speed cameras aim to solve local problems.

How can rural areas be managed to ensure sustainability?

Key ideas	Content detail
• Successful sustainable management requires an understanding of the dynamic nature of social, economic and political processes • Sustainable development of rural areas requires a careful balance of socioeconomic and environmental planning	The study of at least one example to illustrate how planning and management practices enable rural areas to become increasingly sustainable

Key ideas

With reference to at least one example, show how the sustainable development of rural areas requires detailed planning and management of socioeconomic and environmental needs

In MEDCs sustainable use of rural resources focuses on maintenance of services and infrastructure, affordable housing provision and environmental issues. In LEDCs service and infrastructure issues need addressing as do environmental problems such as desertification and land degradation.

Government policies tend to resist **service decline** although the need for economic sustainability can result in some closures. The concentration of services, employment and new housing in key settlements — Framlingham, Suffolk, for example — has been a common management tool.

Larger villages and market towns, together with their spheres of influence, provide the critical thresholds required for functions to flourish. Recently, this policy has been questioned following research suggesting that more mobile rural dwellers do not follow assumed patterns of behaviour. They do not necessarily go to their local service centre but use their increased mobility to visit a variety of destinations for different functions.

Suffolk and Norfolk in the UK are areas where various agencies are attempting to promote sustainable development. The East of England Development Agency (EEDA) and other agencies have a variety of programmes designed to plan and manage change. Commercial organisations can be involved, for example the retailer

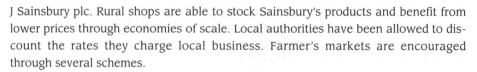

J Sainsbury plc. Rural shops are able to stock Sainsbury's products and benefit from lower prices through economies of scale. Local authorities have been allowed to discount the rates they charge local business. Farmer's markets are encouraged through several schemes.

In LEDCs there are numerous schemes, both small and large scale, seeking to promote rural sustainability. Government and non-governmental organisations (NGOs) such as Christian Aid, Oxfam and Fairtrade, aim to plan for and manage sustainable development projects. The reclamation and plans for management of degraded land, such as the Khushab project, northern Pakistan, represent a deeper appreciation of the need for sustainable issues to be more prominent. Previous over-irrigation led to salinisation of soils and a reduction in yields followed by land abandonment. With more sensitive planning and management, environmental, social and economic sustainability seems possible.

The energy issue

What are the sources of energy and how do they vary in their global pattern?

Key ideas	Content detail
• Energy sources consist of both finite and renewable sources • Energy sources vary in their availability over time and space	The study of the global pattern of energy supply to illustrate that: • The availability of finite and renewable energy sources varies in different parts of the world • That the pattern of energy supply varies over time and space • That an interaction of physical, economic and political reasons is responsible for the pattern of energy supply

Key ideas

What are finite and renewable energy sources?

Principally, fossil fuels (coal, oil and natural gas) are finite: once used they are irreplaceable in timescales meaningful to human society. Renewable energy can be either **recyclable** such as biofuels or hydroelectric power (HEP) or **inexhaustible** such as solar, wind or geothermal power. Currently nuclear energy is based on uranium. Commercial concentrations of uranium are limited although it is found widely in sea water and the Earth's crust. Theoretically uranium could be recycled and used sustainably, but not with current technology.

Finite energy resources cannot be used sustainably as ultimately they will run out. Renewable sources have the potential to be used in sustainable ways.

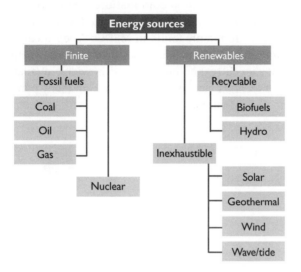

Figure 5 Energy types

Describe and explain the global availability of finite energy sources

Finite sources dominate energy supply at present. At a global scale **coal** has a wide geographical availability with the exception of Latin America and the Middle East. Most coal is consumed in the country where it is produced, with China, USA and India leading the way. It is uneconomic to transport coal over long distances because of its weight and bulk. When burnt, coal has a lower calorific value than oil and gas and there is nothing left once it is burnt. However, there are vast reserves of coal globally with about 150 years' worth of supply at current rates of consumption. Despite declining production, Europe still has considerable reserves of coal.

Oil reserves are less dispersed globally than coal. The Middle East dominates both reserves and production with about 70% of the latter. Oil is traded around the globe in vast quantities: it is relatively easily transported either by pipeline or sea tanker. Areas with limited reserves of their own import oil, Europe and Japan for example. Despite significant domestic production, some countries have such a high demand for oil that they import large quantities, for instance the USA and China.

Natural gas has a wide availability globally. North America and the former USSR dominate production but most continents have substantial production. Reserves are very large and the industry is confident that substantial reserves are yet to be discovered.

Uranium sources are available in most continents, Latin America having the least. However, some 60% is mined in three countries: Canada, Australia and Kazakhstan. Its conversion into electricity production is limited spatially; MEDCs dominate, in particular USA, France and Germany.

Describe and explain the global availability of renewable resources

Of the world's primary energy production, some 13% comes from renewable sources. The vast majority of this energy is from **combustible renewables**, mostly wood used in domestic cooking and heating in LEDCs. HEP, wind, tidal and geothermal power contribute the rest and are mainly found in MEDCs. Their location is down to the development costs of renewables such as wind, solar and tidal energy. These are very high and involve advanced technologies. Wood is widely available and does not require sophisticated technologies to obtain and use. In many parts of the world, its conversion to charcoal for domestic and industrial use is a significant industry.

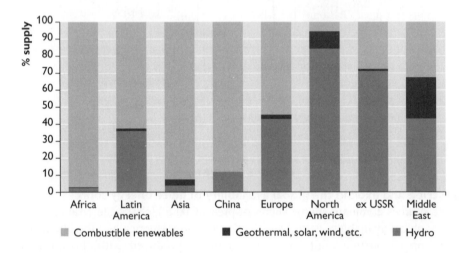

Figure 6 Types of renewable energy by world region

How do physical factors influence the global pattern of energy supply geographically and historically?

Clearly **geology** has a major role to play in the availability of finite sources. Thick coal seams lying close to the surface are the most favoured. The deeper the seams and the more fragmented they are from folding and faulting, the more difficult they are to exploit. Coal also varies in quality depending on its carbon content. Anthracite and coking coals are towards the top end of the scale whereas lignite and brown coals are of low quality.

Oil and natural gas are also affected by geology. Some structures are more favourable for exploitation than others. Pockets of oil and gas can form at the top of **anticlines**. Again, the more disturbed the rocks are by faults, the more difficult it is to access the oil and gas. More recently, technological advances have allowed drilling to be carried out in deep water and rough seas such as the North Sea.

Uranium is quite a common mineral in the Earth's crust but commercial exploitation is restricted spatially. Both open-cast and shaft mining are used although some uranium is recovered as a by-product to copper, silver and gold mining.

Climate has a significant role to play in energy supply. Wind, solar and HEP are closely linked to the right climatic conditions for each. The mining of finite resources is affected by extreme climates. The exploitation of reserves in very cold/hot environments required technological advances before production could begin. For example, drilling for oil in the tundra of Alaska and Siberia is influenced by the environment. Avoiding disturbance to the **permafrost** is a key issue for the energy companies.

How do economic factors influence the global pattern of energy supply geographically and historically?

Economic factors are closely related to how accessible the resources are. Open-cast mining is the most cost effective whereas deep mining via shafts and tunnels is the most expensive. Drilling for oil and gas is less expensive on land than in the sea, particularly if the water is deep. **Freight rates** for transporting energy have tended to fall in the last 30 years. Large-scale ocean-going vessels transport vast quantities of coal, oil and liquefied gas which allow energy-hungry regions to source their supplies from almost anywhere in the world. Likewise these changes allow remote energy producers to operate and reach their markets.

The level of energy prices has much influence on exploration and commercial production of energy resources. Periods of energy shortage give added impetus to the search for new reserves and this can often lead to energy companies operating in more hostile environments. As the first-discovered and exploited reserves come to the end of their viable lives, so less accessible locations are explored and opened up.

Economics have influenced the move to source oil and gas in deep and stormy offshore environments. Deep-mined coal production has been declining in Western Europe and the USA, whereas open-cast methods have been developing, for instance in Australia and China. Extracting uranium from sea water might become a viable proposition if energy prices rise substantially in the future.

The significant increase in interest in renewables such as wind, solar and tidal power is strongly influenced by economics. As their exploitation in terms of electricity production is relatively recent, development and production costs are high and so only affordable in MEDCs. Some large-scale HEP plants are located in LEDCs and there are also many projects using small-scale turbines for single villages or small towns.

How do political factors influence the global pattern of energy supply geographically and historically?

Energy resources have become increasingly politicised, both in their supply and demand. Coal, once the dominant energy source in MEDCs, is no longer seen as being acceptable due to its association with carbon dioxide and acid rain production, and its role in the enhanced greenhouse effect. In some LEDCs, India and China for example, the demand for energy to sustain their developing economies is so high and their domestic supplies of coal so extensive, that politics dictates coal production must proceed unimpeded.

Oil is the energy source probably most influenced by political factors. Oil prices are very sensitive to political events in both oil-producing and oil-consuming regions. **Wars** between countries such as Iraq and Iran and the Iraqi invasion of Kuwait caused steep rises in prices. Tensions between Russia and the USA have an immediate effect on both oil and gas prices.

Because of the dangers posed by nuclear energy and its association with weapon development, the pattern of use of nuclear energy is strongly affected by politics. The oil crisis in the early 1970s led to countries heavily dependent on oil to develop their nuclear programme, France and Japan for example. A variety of regulatory bodies exist to control the nuclear industry, and with growing interest and capability from several countries, nuclear energy will continue to be a highly political issue.

Concerns regarding the links between **climate change** and human activities have meant that interest in renewable sources producing no carbon dioxide have taken on a highly political nature. Having established their high living standards through extensive use of fossil fuels, MEDCs now look to reduce their carbon footprints by using alternative technologies. The use of carbon-emitting energy production by LEDCs to sustain their development is a sensitive political issue.

What is the relationship between energy use and economic development?

Key ideas	Content detail
As economies develop, the demand for energy increases	The study of the global pattern of energy use in relation to economic development. Included in this is a statistical investigation
• There are significant differences in energy use between MEDCs and LEDCs • There are significant differences in energy mix between MEDCs and LEDCs	A study of two contrasting countries from opposite ends of the development spectrum, Sweden and India for example, to illustrate their different energy use and mix of energy sources

Key ideas

How and why does energy use change with economic development?

Energy is needed to produce all goods and services. The more that are produced and consumed, the more energy is required. High levels of production and consumption generate economic development. With wealth comes the ability to produce and consume at high levels. Because of their poverty, millions living in LEDCs are only able to consume and produce locally. Agriculture, especially subsistence production, uses human labour and animals; cooking and heating in homes rely on biofuels, mostly wood, and manufactured goods tend to come from small-scale producers who use only small quantities of energy. Mass consumption of manufactured goods which require vast amounts of energy to produce is a feature of MEDCs. Lifestyles in MEDCs are high in energy consumption in terms of transport, home and work-place heating and cooling, and leisure.

What is the strength of the relationship between energy use and level of development?

The relationship between economic development and energy consumption is a strong and positive one. The two factors increase hand in hand, and are both cause and effect. However, there are anomalies from the overall pattern. Energy-rich nations, such as some of the oil producers, have high energy usage due to cheap oil and relatively low GDP per capita.

How and why do countries vary in their use and mix of energy?

Energy mix is the combination of energy types that makes up the energy supply within a country. Four factors interact to influence a country's energy mix.

- The **energy resources** available domestically are usually cheaper than imported ones. India and China have vast reserves of coal so rely on this for much of their energy; Saudi Arabia possesses significant oil reserves and so predominantly uses this type of energy.
- **Energy security** is a growing concern for many countries. Having a diversified energy mix reduces a country's dependency on any one energy source. This can be important strategically where energy comes from politically unstable regions, oil from the Middle East for example.
- The **level of development** is an influential factor. MEDCs use advanced technologies in their energy industries so fossil fuels, nuclear and some renewables make up their mix. LEDCS are reliant on traditional energy sources such as biofuels, wood and animal dung.
- **Environmental concerns** are increasingly influential amongst MEDCs. Reducing fossil-fuel use is seen as critical in achieving reductions in greenhouse gas emissions and so lessening global warming. More emphasis is placed on the role of renewables such as wind power, and nuclear energy is receiving more attention.

What are the social, economic and environmental issues associated with the increasing demand for energy?

Key ideas	Content detail
The exploitation of energy resources brings both opportunities and problems for people and the environment	A study of two contrasting examples (e.g. Alaska and the Niger Delta) to illustrate the social, economic and environmental opportunities and problems created by energy resource exploitation

Key ideas

Show how energy resource exploitation can bring social and economic opportunities

Energy use is clearly linked with economic and social development. Many countries try to secure improved energy supplies in order to create and sustain development both nationally and locally. Major energy projects such as the Three Gorges Dam, China, or the Itaipu Dam on the Brazil–Paraguay border generate such vast quantities of electricity that they are designed to aid development on a national scale. In particular they are planned to sustain the economic opportunities of core regions such as the Yangtze Delta and Shanghai in China and the states of Minas Gerais and Sao Paulo in southeast Brazil. Exploitation of oil reserves on the North Slope of Alaska has brought economic advantages both to the state of Alaska and to the US economy as a whole. Improved energy security, reduction in imported energy and a growth in employment in oil industries are all seen as valuable opportunities. The Dinorwig pump storage scheme in Snowdonia has helped boost the local economy and offers opportunities in a locality where few exist.

What problems arise from the exploitation of energy resources?

There are a variety of 'costs' associated with the exploitation of all energy resources, non-renewable and renewable.

Environmental degradation is common to most schemes. Mining and quarrying produce waste or spoil which is dumped. If not managed correctly this waste can prove hazardous, for example by polluting water courses or becoming unstable, leading to collapse. At Aberfan in South Wales a coal waste tip became unstable due to a lack of managed drainage. It flowed down into the small town burying some houses and the school which was in session. The death toll was 144, of whom 116 were children. Oil leaks, at any stage of the production process, can be highly damaging, especially if they occur in remote and fragile environments such as northern Alaska and Siberia. The transport of oil and nuclear material is especially hazardous. Tanker accidents can spill vast quantities of oil into marine and coastal environments. The ecological impact is severe, as it is with accidents leading to the release of radioactivity from nuclear power stations.

Visual intrusion into landscapes is common with nearly all types of energy production. Power stations tend to be large buildings; dams block valleys and their reservoirs flood areas upstream; wind turbines are sited in exposed locations to catch as much wind as possible; tidal barrages cut across estuaries.

Social sustainability can be adversely affected by energy projects. Dam construction often requires resettlement of people living in areas flooded by the reservoir. Cultural, archaeological and historic sites can be lost under water or in the course of mining. Traditional ways of life can be disturbed as in northern Alaska. In the Niger Delta, west Africa, where there are substantial reserves of oil and gas, the oil industry has come into frequent conflict with indigenous people. Traditional ways of life depended on fishing, subsistence farming and using the tropical forests for food and materials, all of which have been adversely affected by the oil industry.

Although energy projects have economic advantages, they also have economic costs. Reservoirs claim land that was once productive farmland. Less fertile alluvium is carried by the river as silt is trapped behind the dam. Farmers below the dam then have to use costly chemical fertilisers. The nutrients in the silt are also denied to estuary ecosystems, often having an adverse impact on inshore fishing.

How can energy supply be managed to ensure sustainability?

Key ideas	Content detail
Managing energy supply is often about balancing socioeconomic and environmental needs. This requires detailed planning and management.	The study of at least one example (e.g. California) to illustrate how energy demand can be planned and managed in sustainable ways

Key ideas

For a named example, explain how planning and management are helping to satisfy energy demand in increasingly sustainable ways

Finite resources cannot be managed in a truly sustainable way. They can, however, be managed so as to reduce the rate at which they are consumed. Development of renewables can be promoted, although even with these there are costs.

California is perceived as an energy-intensive location. Its large population, nearly 37 million inhabitants, and substantial economy, comparable in size to the UK's, have very large energy demands. Some of the highest living standards on Earth are found in California which demand much energy to support. Until recently most of the energy supply came from fossil fuels, namely oil and gas. California has local supplies of both.

The last three decades have witnessed something of a revolution in attitudes and approach in terms of energy supply and demand in California. Electricity consumption per capita has hardly altered but the state's carbon dioxide emissions have fallen by about a third.

California's planning and management has focused on transport, electricity production, energy efficiency and development of renewables. Los Angeles in particular and California in general have high car ownership and usage. State and local authorities have introduced tough laws aimed at cutting vehicle exhaust emissions. Vehicle manufacturers must introduce cleaner technologies and motorists are encouraged to convert to non-petroleum fuels such as biodiesel, liquid natural gas (LNG), dual-fuel or completely electric vehicles.

The electricity generators have been managed in ways to encourage them to use cleaner technologies. Coal-fired power stations, which emit high levels of carbon dioxide, are at financial disadvantages compared to gas-fired installations which emit fewer greenhouse gases. Initiatives such as a levy on greenhouse emissions aim to encourage the generators to invest in renewables. California generates more electricity than any other state from renewables. There are some 14 000 wind turbines in the state but many are now ageing and need replacing. Expansion of wind farms has environmental costs, such as the land used and impacts on fragile desert environments and ecosystems. Solar power used by both power stations and private households is encouraged. California receives large amounts of insolation and there are plans to expand solar power considerably. The mountains of the Sierra Nevada are home to some HEP plants, most of which have been in existence for many decades. They have significant environmental costs attached and it is unlikely that this energy source will be expanded. The location of California on a tectonically active region has meant that increasing attention is being given to the potential of geothermal energy. Hot water and steam from underground is used to generate about 5% of the state's electricity. Hot water from underground is also being used to heat homes and greenhouses.

Planners have also tackled the energy issues associated with demand from homes. Standards of energy efficiency have been raised so that new homes require much less electricity both to heat and cool than older ones.

The growth of tourism

In what ways has the global pattern of tourism changed?

Key ideas	Content detail
• Tourism has developed into a global industry and now features in every continent • A variety of factors influence the growth of tourism and its changing patterns These factors vary from place to place	The study of the global pattern of the growth of tourism to illustrate: • Changes in the location of tourism • Changes in the type of tourism • The social, economic and political reasons behind the growth of global tourism

Key questions

What is meant by tourism?

The World Tourism Organization's definition is '...the activities of persons travelling to, and staying in, places outside their usual environment for leisure, business and other purposes'.

Describe the different types of tourism

It is possible to classify tourism in several different ways. Factors such as purpose (e.g. business/recreation), location (e.g. beach/urban), scale (e.g. large/small) and sustainability (e.g. mass/ecotourism) can help distinguish tourism types.

One key distinction can be made between **mass tourism** and **ecotourism**. The former is seen in large-scale beach resorts where interconnections between travel agents, airlines, tour operators and hotel chains 'manufacture' holidays for thousands of tourists. Stretches of the Mediterranean and the east coast of the USA possess many of these resorts.

Alternatively, ecotourism caters for small numbers of visitors focused on specialist activities such as viewing wildlife or historical sites and visiting remote locations such as the Himalayas.

Enclaves are tourist locations largely separated from their surroundings. Commonly found in LEDCs, they are used by MEDC visitors who have little engagement with the local community.

A wide variety of tourism types have recently emerged that include visits to urban locations, ocean cruising, visits associated with sport as either spectators or participants and activity holidays.

What are the trends in global tourism?

There has been steady average growth of 6.5% per year in international tourist numbers since the 1950s. This trend has persisted through various economic and political crises although with some periods of particular growth or decline within it.

Tourism has also expanded geographically. Both as sources of visitors and as destinations, more and more countries have become engaged in tourism. **Globalisation**, the process of integration of economic, social and political systems around the world, is clearly evident in tourism.

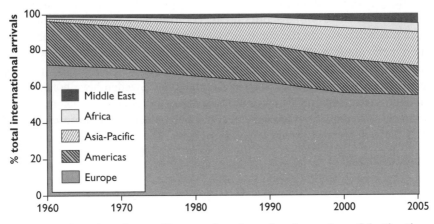

Figure 7 Changing share of international tourism by region of destination, 1960–2005

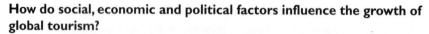

How do social, economic and political factors influence the growth of global tourism?

Increases in **paid holidays** have allowed people in MEDCs more opportunity to travel. They receive several weeks of time off work, which allows longer journeys to be made to more distant holiday destinations. The fact that during this time off people are entitled to receive pay means that they can afford to go away and travel further.

Education to higher levels encourages more international tourism as people become familiar and competent with foreign languages; this applies both to tourists and those in the industry. Education also encourages people to be more aware of possible destinations and the variety of resources (physical, cultural and heritage) that tourists can use. Increasingly effective forms of communication, in particular the internet, allow tourists to remain in touch with their home area and to contact potential destinations to make travel and accommodation arrangements. This encourages people to travel internationally. Overseas access to credit and cash is an advantage of many banks being **transnational corporations (TNCs)**.

The ease with which people can obtain money via electronic transfers using automatic telling machines (ATMs) is another factor. Hotel chains have also become international and so offer standardised facilities and familiar food, making it easier for some people to go abroad. The positive reputation of many chains reassures people to stay in one of their hotels, especially if the destination is in an unfamiliar culture. Tour operators take charge of all arrangements for certain types of holidays. People can simply visit one agency to book travel and accommodation and to receive expert advice regarding practicalities such as visas and health matters — for example, which inoculations are necessary.

In MEDCs people's **disposable income** has risen. This has given them more money to spend on holidays after paying for essentials such as housing, food, clothing and taxes. Many people are able to take more than one holiday a year — a principal period away combined with a shorter break. For example a family might take a fortnight in a beach resort in the summer and a long weekend in an historical urban location in the autumn. The increase in wealth has been sufficient to allow many of the better off in MEDCs to own holiday homes in different countries. In Europe, people from northern countries such as Germany and Sweden have holiday homes in Mediterranean regions or mountainous areas for beach and skiing holidays respectively. Affordability of foreign travel has been aided by **economies of scale**, such as those achieved through larger aeroplanes. The absence of tax on aviation fuel has, in part, assisted airlines, in particular the growth in low-cost carriers.

Since the 1950s the absence of major international conflicts has helped stimulate global tourism. Short-term impacts from some **geopolitical incidents** have led to falls in tourist numbers but recovery has tended to be relatively quick. Instability in some regions and individual countries holds back tourist developments. Supranational political organisations like the EU aid tourism through developments such as a common currency, the euro, and relaxing border controls on movement of people. Many LEDC governments actively promote a tourist industry so as to attract MEDC visitors. Wealth the tourists bring in can then be used to fund development projects.

What is the relationship between the growth in tourism and economic development?

Key ideas	Content detail
As economies develop there is an increased demand for tourism	• The study of the global pattern of tourism in relation to economic development. Included in this is a statistical investigation • A study of two contrasting countries from opposite ends of the development spectrum, Thailand and Iceland for example, to illustrate how tourism can play a significant part in the economic development of countries and regions

Key ideas
What is the relationship between economic development and tourism?
In general the higher the level of development of a country, the greater the demand for holidays from its population. The demand for tourism increases very rapidly as countries become highly developed, as can be seen in western Europe.

In terms of tourist destinations, there is also a positive relationship between level of development and tourist arrivals.

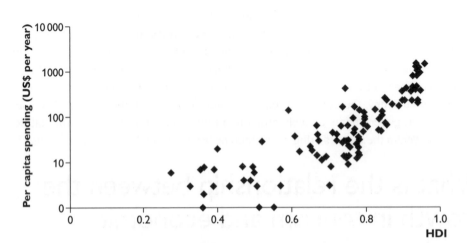

Figure 8 Economic development and international tourist expenditure per capita, US$, 2005

Why is there a relationship between tourism and economic development?

Increasing levels of development are usually associated with improved economic and social conditions as registered by human development indexes (HDIs). Better economic and social conditions allow disposable incomes to rise, people to be more aware of potential holiday destinations through higher educational standards, and practical issues such as vaccinations to be taken care of. Although international tourism is dominated by the world's most developed countries, some LEDCs are experiencing rapid growth as origins of tourists. As the economies of China and India strengthen, the rapid growth in the middle classes is resulting in many millions of people having the disposable income to spend on travel.

Countries without the expected standards of physical infrastructure such as water supply, power and transport, attract only small numbers of 'adventure' tourists. Some LEDCs are, however, showing significant increases in their tourist industries as they seek to invest in facilities designed to appeal to people from MEDCs. Costa Rica and Vietnam are examples of this.

Some countries with high HDIs do not contribute large numbers of international tourists. They might be geographically relatively isolated such as New Zealand. Equally, some countries have so many opportunities for tourism within their borders that their inhabitants tend not to travel abroad, such as the USA.

With reference to contrasting countries, show how tourism can lead to economic development

Some countries explicitly seek to exploit their tourism resources in order to help stimulate economic development. It is not just LEDCs such as Thailand that engage in such programmes. Many MEDCs have recognised the value of attracting tourists, for example the UK and Iceland.

The economic advantages can be seen in the **wealth creation** that tourism encourages. Tourists need to use local currencies, so they change their money. If they come from an MEDC then their domestic currency (e.g. dollar, euro, pound) is often perceived as being 'hard', that is of greater worth internationally than the 'soft' currency of an LEDC (e.g. shilling in Kenya or baht in Thailand). Even MEDCs need to earn **foreign currency** so as to avoid having a very uneven balance of trade as all countries spend money overseas on imports. Iceland, for example has a very small manufacturing sector and needs to pay for the import of vehicles and the fuel required to keep them running.

Jobs in tourism generate **wages** which create wealth both for individuals and nations. Taxes can be levied which can then be spent on programmes such as education and health care.

Indirect development can arise from **infrastructure** developments that are primarily aimed at tourists. Roads, for example, might be constructed to aid tourist access to a region but as a consequence local people can use them to transport agricultural produce to markets. Improvements in the supply of utilities such as water and power to tourist developments might extend to local communities. Such projects can attract **Foreign Direct Investment (FDI)** which can complement government spending, which in the case of many LEDCs is, of necessity, small scale.

What are the social, economic and environmental issues associated with the growth of tourism?

Key ideas	Content detail
The exploitation of tourist resources brings both opportunities and problems for people and the environment	A study of two contrasting examples (e.g. St Kitts and Nevis and Tenerife) to illustrate the social, economic and environmental opportunities and problems created by tourist resource exploitation

Key ideas

Show how the growth of tourism can bring economic and social opportunities

Tourism is a **labour-intensive** industry. Direct employment in tourism can be substantial in both LEDCs and MEDCs, for example within travel companies and hotels. Indirect employment results from the **multiplier effect**. This describes how jobs in transport, agriculture and manufacturing can be linked with tourism via the services and products they generate, some of which are used by tourists. Those directly employed in tourism have wages to spend, so in turn create a demand for goods and services that other locals can supply, and so the effect continues.

Both direct and indirect employment can be significant in areas where there are few other opportunities. Caribbean islands such as St Kitts and Nevis have high levels of unemployment due to the collapse of their sugar industries and few alternative sources of income. Regions in MEDCs such as the Lake District and Cornwall turn to tourism for jobs as traditional local sectors such as agriculture and mining decline.

Hand in hand with economic development come social opportunities. Rather than suffer unemployment or underemployment in jobs such as farming, people occupied full time in tourism can let their children attend school. They can also afford better housing which is likely to lead to improved health. There are usually many opportunities for female employment in tourism which might help enhance the role and status of women. In some rural regions in MEDCs where tourism replaces traditional activities, people do not migrate away and so communities stay together.

What environmental opportunities can tourism bring?

If tourism develops to the point that significant income is earned from it, then local communities become keen to protect and conserve the natural environment. Developments are scrutinised to make sure that they are as sustainable as possible. **Environmental impact assessments** are carried out to make sure that the very resources that are attracting people are not destroyed. On St Kitts rainforest has been given legal protection and building heights limited to minimise the effects of development.

Some countries use a system of national parks to manage visitor impacts. On Tenerife Mount Teide is not only protected by Spanish law but it is also now designated as a World Heritage Site by UNESCO.

What problems arise from the exploitation of tourist resources?

Environmental impacts can be severe in some locations. Poor planning, either from the absence of rules or a failure to enforce them, leads to inappropriate development. Both in terms of location and scale, building can severely detract from the beauty of a site as developers want to exploit the best views and positions for hotels. **High-rise development** has occurred along many coastal zones in both LEDCs and MEDCs. In some places even access to the beach and sea has been interfered with by buildings. Tourist activities can harm local environments. Coral reefs are easily damaged by divers breaking off pieces as souvenirs and by boat anchors dragging across their surface. Oil spills from cruise liners are serious in regions such as the Caribbean. Offshore dredging to replenish sandy beaches is destructive of sea-bed ecosystems.

Two major concerns common to many tourist destinations are those of **water supply** and **sewage disposal**. Per capita demand for water is usually higher from tourists than locals. Over-abstraction of groundwater and river water can cause shortages and environmental damage to ecosystems. Tourist-related activities such as golf consume vast volumes of water. The treatment and disposal of waste water presents difficulties. Without effective treatment, foul water enters streams and rivers and then the sea. Ecosystems can be badly affected, for example inshore fisheries and coral reefs.

Economic impacts can be negative. As international tourism has grown, so has the role of tourist-focused TNCs. Capital-intensive developments owned by TNCs, such as resorts and large hotels, tend to result in revenue flowing back to the TNC's country of origin. This **leakage** reduces the wealth entering local economies and diminishes the multiplier effect. Leakage is at its most extreme from enclaves. Additionally tour operators, international airlines and insurance companies tend to be based in MEDCs, increasing the proportion of holiday profits that divert from the destination. In some locations food is flown in rather than being sourced locally. Tourism is often seasonal resulting in unemployment during the slack months. It is also an industry very susceptible to downturns in the global economy, to changes in fashion, that is which tourist destinations are 'in' and which are 'out', and geopolitical events.

Social impacts can be seen in the disruption to local communities caused by the influx of people from different backgrounds and cultures. Locals abandon agriculture and young people give up education with the chance to make 'easy' money in tourism. When tourism numbers decline, these people and their communities are left with few alternatives. The bringing together of very different cultures can generate tension, as may occur when the wealth of MEDC visitors is displayed to LEDC hosts. Crime of various sorts can spring up, some of it deliberately encouraged by the demands of MEDC tourists for drugs or prostitutes.

How can tourism be managed to ensure sustainability?

Key ideas	Content detail
Managing tourism to ensure sustainability is about balancing socioeconomic and environmental needs. This requires detailed planning and management	The study of at least one example (e.g. the English Lake District, Arches National Park, USA, or Annapurna, Nepal) to illustrate how sustainable tourism, including ecotourism, operates in conjunction with communities and the environment

Key ideas

For a named example, explain how planning and management are helping to achieve sustainable tourism

Sustainable tourism attempts to meet the needs of a present-day tourist industry without compromising the well-being of future generations. That well-being can be seen in physical, economic and social terms. Increasing attention is given to the management of tourism both in MEDCs and LEDCs.

Tourism uses a range of resources. Where these are physical, care is taken to ensure that their quality is maintained. The impact on local communities is also in need of management so that communities can flourish without relying solely on tourism. The traditions, cultures and customs of such communities need the assistance management can provide. National parks are used by many countries to assist such management.

The English Lake District in northwest England emerged as a popular tourist destination during the nineteenth century. The region, with its dramatic exposed upland glaciated scenery of fells, peaks, valleys and lakes, attracted artists and romantic poets. Additionally the human landscape of small market towns, villages and isolated farms with their vernacular architecture and dry stone walls enhanced the region. Pictures and writings stimulated interest and with the development in transport technology of first the railway and then the internal combustion engine, increasing numbers of people were able to experience for themselves the resources of the Lake District. By the middle of the twentieth century it was clear that more formal management was required to sustain the Lake District, both as a tourist destination, but also as a region where communities live and work.

Some key issues of sustainable management have emerged in the Lake District. One impact of increased numbers of tourists is increased road traffic. Traffic management schemes propose the use of shuttle buses along the most popular routes (not yet introduced), limiting parking spaces and raising parking charges. Conflicts between different groups of recreational users in water sports and angling have been addressed by the introduction of a 10 mph speed limit on Lake Windermere. Footpath restoration is used to counter erosion along the more popular tracks. The park authority is also active in encouraging improvements in the quality of accommodation and retailing in towns. Planning rules are applied to ensure that building styles match local architecture and use local materials. Changes in farming practices and the various influences on agriculture, such as the level of grants and subsidies, create pressures on landscape quality. Both the EU and DEFRA have a range of policies designed to encourage sustainable farming and help conserve the environmental value of landscape. Afforestation is now more controlled than in the past so that woodland fits appropriately into the landscape.

Questions
&
Answers

This section contains examples of student answers to eight examination questions, covering the four topic areas outlined in the **Content Guidance** section of this guide — managing urban change, managing rural change, the energy issue, the growth of tourism.

Section A contains four structured data-response questions (one for each option), and Section B has four extended-writing or essay questions (again, there is one question for each option).

Each structured data-response question consists of four sub-parts and is constructed around stimulus materials such as OS maps, sketch maps, diagrams, charts and photographs.

In the examination, you will have to answer two four-part structured questions and one extended-writing question. You should allow yourself around 30 minutes for each question. The lengths of your answers to the structured questions should be proportional to the mark weighting, which varies from 4 to 9 marks. A rough guide is to think in terms of 1 mark for every 1.5 to 2 lines of writing.

Examiner's comments

Each student answer in this section is followed by **examiner's comments**, indicated by the icon ℮. These comments show how marks have been awarded and highlight areas of credit and weakness. For weaker answers the comments suggest areas for improvement, by highlighting specific problems and common errors such as lack of development, excessive generalisation and irrelevance.

Section A: structured questions

Managing urban change

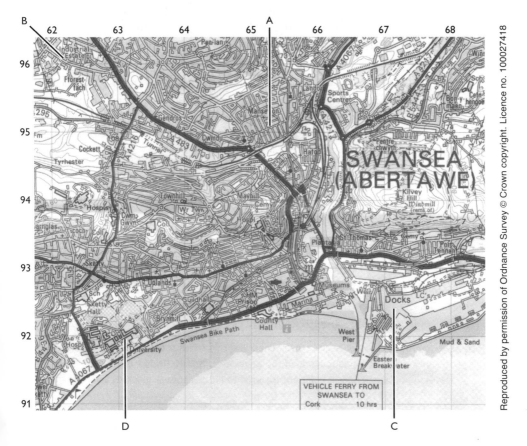

Figure 1 OS map of part of Swansea

(a) Study Figure 1, an extract of an Ordnance Survey map showing part of Swansea.

   (i) Identify the four different types of urban land use at locations A to D
       on Figure 1.   (4 marks)

   (ii) For any **ONE** of these land uses, state and explain two reasons why it
       is taking place in an urban location.   (6 marks)

(b) Outline two social issues resulting from urban change.   (6 marks)

(c) With reference to one or more located urban areas explain methods used
    to manage air pollution.   (9 marks)

◼◼◼

1

question

C-grade answer

(a) (i) A = housing; B = industry; C = docks; D = university   4/4 marks

🖉 All four have been correctly identified.

(ii) Industry is often located in urban areas as this is where the industry was developed. Factories need labour and there are many people living in urban areas who need jobs. Factories also need raw materials and locate where they can have easy access to them.   2/6 marks

🖉 This is a level 1 answer even though is does offer two reasons. The first one concerning labour supply is valid although not well expressed. The comment about raw materials is not specifically about 'urban' locations. This point has potential if the suggestion is made that focusing transport routes on urban areas allows raw materials to be transported in from a variety of locations.

(b) One type of urban change is inner-city redevelopment. In some places such as Birmingham this has meant a lot of redevelopment. Many of the nineteenth century terraced houses have been demolished as they were not fit to live in. However, they have been replaced by tower blocks and these have not been good places to live in either. The sense of community has been destroyed and people don't get on with each other.

Another social issue has been the building of housing estates on the edge of cities. This has meant that some families have been split up. It also means that these people have long journeys to the CBD to find work or to shop.   5/6 marks

🖉 Two appropriate social issues are outlined and related to urban change. The candidate shows a clear understanding of the issues even though his/her expression lacks precision in places. There is sufficient material to lift this answer just into level 2.

(c) Air pollution is a major concern in some urban areas such as Santiago. It comes from burning fossil fuels which releases harmful gases such as carbon dioxide and sulphur dioxide as well helping create smog. The growth in traffic contributes heavily to air pollution as very few cars have catalytic converters which reduce pollution. Santiago's population has been growing rapidly which has meant the city has enlarged so that longer journeys take place. The government is trying to sort out the bus system which is chaotic. They are building bus lanes to speed buses up so that more people will use them and not their cars. Many of the buses are old and so their engines are inefficient. The government is encouraging the bus companies to buy new buses that are more efficient. They are also building a metro system to carry more people so that car use will reduce. There is also the building of motorways so that traffic flows will improve. Cars that keep moving cause less air pollution than if they are stuck in a traffic jam.   5/9 marks

This answer begins by describing the causes of air pollution which is not required by the question. This wastes valuable time in the exam. Once the candidate starts discussing the methods being used to reduce air pollution, the response becomes more convincing. All the methods mentioned are valid but to lift this answer to the top of level 2 or into level 3 more place detail about Santiago is required. The answer is therefore at the bottom of level 2.

The total score for this four-part answer is 16/25, a secure C grade.

■ ■ ■

A-grade answer

(a) (i) A = housing; B = industry; C = docks/industry; D = university   4/4 marks

All four have been correctly identified.

(ii) Housing dominates the land use in urban areas. As towns grew, more and more people needed somewhere to live. People moved to towns as there were many jobs created in the industries setting up in the nineteenth century. Large areas of terraced housing were built to house workers as close as possible to the factories as personal mobility was low. By the end of the twentieth century services were more important and these were concentrated in urban areas so housing was still needed. Suburbs developed as people had cars and were able to commute.   5/6 marks

The opening sentence is an encouraging start to this level 2 response. The historical perspective is valuable and gives a depth to the answer. In the second reason, about the development of services, it is not clear whether the answer is discussing employment in or use of services. Two distinct paragraphs would help here.

(b) There is much urban change occurring in LEDC cities. They are growing rapidly through rural–urban migration and natural increase. Kenya's urban population is growing at about 6% per year. In Nairobi there are vast areas of slum housing such as Kibera. The inhabitants cannot afford to pay for any better accommodation and so have no choice. These houses are often poorly built with no clean water or toilet facilities. There is much overcrowding.

Another social issue is the problem of unemployment. Many of them have no regular work and just make do with unskilled casual work. Many work in the informal sector selling things on the streets but this does not bring them very much money.   5/6 marks

Two issues are clearly identified and related to the question. The context is well described at the start of the answer, namely the rapid growth in urban population. The comments about housing and unemployment are effective but perhaps could be strengthened with the use of some figures: for example, population densities in Kibera can reach 90 000 persons per square kilometre. Nevertheless this is a good level 2 response.

(c) Air pollution has been an issue in urban areas for a long time. In the nineteenth century the use of coal to heat homes and for industry to generate steam power meant there was much air pollution. In London smog was common in the winter up until the 1950s. Then a Clean Air Act was passed which banned coal burning unless smokeless fuel was used. The air quality improved and smog is now very uncommon.

However, with the growth in traffic cities such as London suffer from air pollution from exhaust fumes. Nitrous oxides, carbon monoxide, carbon dioxide and particulates are emitted and cause health problems for many people. London has introduced a congestion charge in the central areas which has helped improve air quality. In Los Angeles they also have problems from exhaust fumes. The gases interact with the sunlight to form ozone and a photo-chemical smog which is a health hazard. They try to manage the traffic such as with car sharing lanes on the freeways to encourage people to share lifts.

Some LEDC cities suffer badly from air pollution. Mexico City is in a high altitude basin which traps air pollution in an inversion. The government is trying to stop people using their cars so much and to phase out the older inefficient buses as they emit so much particulate pollution.   9/9 marks

🖉 This is a successful level 3 response that gives information on several valid types of management. The use of real-world examples, although quite brief, strengthens the answer. It is important that the focus is on the management of air pollution and in places a little more detail could be included. For example, the comment about Mexico City and car use is vague. But overall this answer is well written with an effective structure, uses terminology appropriately and is convincing.

🖉 **The total score for this four-part answer is 23/25, a secure A grade.**

Managing rural change

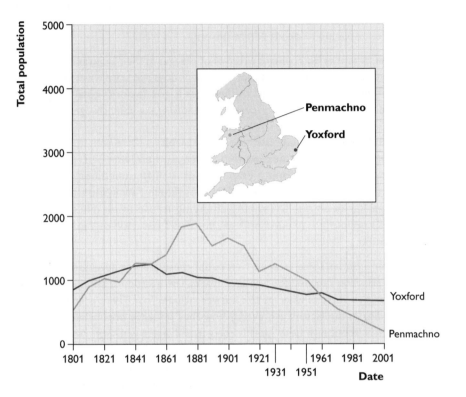

Figure 2 Population change in the rural settlements of Yoxford and Penmachno, 1802–2001

(a) Study Figure 2, population change in the rural settlements of Yoxford and Penmachno.
   (i)   Describe the changes in population for both villages.   (4 marks)
   (ii) Suggest **TWO** reasons for these changes.   (6 marks)
(b) Show how the development of rural areas can be influenced by governments. (6 marks)
(c) With reference to one or more located rural areas, explain how land-use change can bring about land degradation.   (9 marks)

■ ■ ■

2

question

D-grade answer

(a) (i) Yoxford is at nearly 1000 in 1801. It increases to just over 1000 but then is at less than 1000 in 2001. Penmachno is at 500 in 1801 but then becomes nearly 2000 before reducing to about 250 at the end.   **2/4 marks**

e The answer describes both villages in terms of changes in their population totals. However, it lacks attention to detail, for example it does not state when the populations of the two villages peaked, neither does it give Yoxford's total in 2001. The answer is scored as level 1 for correctly identifying the basic changes.

(ii) There will have been a high birth rate to start with as there was no contraception. Families would have needed to be large as there was high death rates. Also there was lots of farming employment as little machinery was used. There was a decline in the village populations due to two world wars. Most of the change has been due to farming changing to use more machinery. This means less people are needed and so they don't have jobs. There is a lot of out-migration.   **3/6 marks**

e The answer includes some relevant factors but these are not developed to the level of explanation required for high-level marks. The question is also clear in its demand for two reasons whereas this answer includes four. There is insufficient understanding of the factors and exactly how they influence population change. The comment about high birth rates is correct, and linked with high death rates, has some merit. No timescale is put on this, however, so the examiner is left to work out which part of the trend this comment applies to. The comment about changes in farming practices and the demand for labour is basically correct but not allocated to a time period. Overall there is limited appreciation of the reasons.

(b) Governments are important in rural areas. Agriculture is influenced by governments. They can set rules that farmers have to follow such as grants and subsidies. If they produce too much then governments have to buy the produce such as milk or cereal. Governments can also pay for drainage of fields to improve soils. In some places farmers are paid to look after the landscape which means they rebuild walls and hedges.   **2/6 marks**

e The answer focuses exclusively on farming but does link this with government. Although the candidate correctly identifies several relevant influences, the link between government and rural development is vague. No details are given and some of the comments are out of date. Whenever government influences are studied, it is important to be as up to date as possible. This is a level 1 answer.

(c) In East Anglia there has been a lot of land-use change. Farmers have been taking out hedgerows in order to make their fields larger. They do this to use large-scale machinery such as combine harvesters so that they can grow cereal crops. The fertilisers they use on the crops can run off into the rivers and this reduces the water quality. Algal blooms can grow and this destroys the food chains. Pesticides are used a lot by farmers and these interfere with birds. There are fewer insects for birds to eat and some pesticides end up in the shells of eggs making them too thin so that they break before hatching. Pesticides go up through the food chain to the top carnivores.

When hedgerows are removed wind erosion can take place. This is when the soil is bare often in winter so the strong easterly winds pick up small fragments of soil. If there is a lot of rain, water can wash away the soil as well. Sandy soils are mainly the ones affected by this erosion.

Some towns have grown in East Anglia such as Norwich and Thetford. New houses and industry have been built on greenfield sites. This has meant that rural ecosystems have been replaced by urban ones and this has led to land degradation. Ecosystems have been destroyed but in some areas they have set up nature reserves. In the Broads there are rules about using boats so that their speeds are restricted to prevent banks being washed away.   5/9 marks

✍ Although the candidate makes several valid points, none is developed in full or with sufficiently convincing explanation to be secure. It is not clear that the candidate fully understands the term 'land degradation' and a concise definition would help tie the answer closely to the question. Comments about changes in farming are valid but the actual changes are implied rather than explicitly stated. The link between land-use change and land degradation needs to be clearly expressed so that cause and effect are evident. This is a level 2 answer, which with a sharper focus on the actual question and tighter prose, would achieve level 3.

✍ **The total score for this four-part answer is 12/25, a D grade.**

■ ■ ■

A-grade answer

(a) (i) Both villages increase during the nineteenth century but then decline during the twentieth. Penmachno has more growth to nearly 2000 in 1881 but then declines to below Yoxford in 2001, 200 compared to 700. Yoxford is a smaller village but has not declined as much.   4/4 marks

✍ The answer opens with an accurate summary statement. It then gives some appropriate factual detail for both villages. This is a level 2 answer.

(ii) The growth in the nineteenth century was due to natural increase. Also farming was doing well with more food required by a growing population in urban areas. During the twentieth century there is out-migration, especially of young adults which means there are fewer births in rural areas. By the end of the century, Yoxford and Penmachno were mainly retired people.

2

question

More labour was needed in farming to start with but this declined with mechanisation. Also rural-urban migration took place as urban opportunities were better. Rural services have declined as their threshold declined as people left villages. Penmachno is a remote rural area in North Wales and very few people want to live there as there are few jobs.

5/6 marks

The answer targets the population changes but it lacks a really sharp focus on two reasons. The reasons offered are valid and the candidate's understanding of them is secure. It is encouraging that correct terminology is used, 'threshold' for example. This is a level 2 response.

(b) Governments have had major influences on rural areas. In the EU the Common Agricultural Policy (CAP) has given grants and subsidies to farmers. In the past this has encouraged farmers to change what crops they grow. For example in parts of East Anglia many farms specialise in cereals and break crops such as oilseed rape. This is because farmers received bigger payments for these crops. Now the Environmental Stewardship scheme pays farmers if they farm sustainably for example leaving strips of land uncultivated and reducing fertiliser use.

With rural depopulation in many areas governments have tried to stop decline. Some rural areas have a key settlement policy when local authorities such as county councils work with DEFRA to help sustain local communities. Funding is given to support services such as health centres and local shops. Planners try to make sure housing is available for local residents which is affordable. This helps younger families stay in the area.

6/6 marks

The key command here is 'show how' which means a clear link is required. The candidate does this successfully with two different examples: this is a sensible strategy as it avoids the risk of repetition. The use of two paragraphs illustrates the effective planning of the answer, even though there is limited time. Both parts include relevant material and display secure understanding so that this is a clear level 2 response.

(c) Land degradation is the reduction in the quality of land, usually for agriculture. It is caused by several factors such as soil erosion and salinisation.

There have been significant changes in agriculture in parts of the UK. Farmers have been encouraged to grow more cereals by the grants and subsidies paid by the CAP. In order to use large machinery which makes this type of farming so efficient, hedgerows have been removed. This has left huge fields with no windbreaks and so when the soil is bare before the new crops have grown, strong winds can cause wind erosion of the top soil. Also the exposed soil can be washed away by heavy rainfall over winter. This has happened in parts of the South Downs.

In Nara, Mali, there has been land degradation. The low rainfall, less than 500 mm per annum means that drought is common. Rainfall is also very irregular which makes it difficult for the local people to grow crops and keep livestock. Because of population growth, more food is needed and so tribes such as the Fulani wanted to increase their herds. This led to overgrazing of the desert vegetation. They also needed more wood for firewood. Strong winds picked the unprotected soil and carried it away as dust storms. When it does rain in the Sahel, there are usually heavy rainstorms which wash away soil that is not protected by vegetation.

8/9 marks

e This is a level 3 answer: knowledge and understanding of land degradation in two different locations is quite detailed. The link between land-use change and degradation is explicit in each of the main paragraphs. An authoritative tone is set with the accurate definition at the start. The candidate could have given a little more place detail on the South Downs example, but overall this is a convincing piece of extended writing.

e **The total score for this four-part answer is 23/25, a clear A grade.**

questions & answers

Question 3

The energy issue

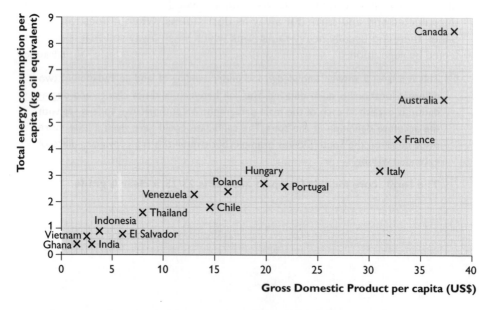

Figure 3 Relationship between Gross Domestic Product and total energy consumption, both per capita, for selected countries, 2007

(a) (i) Describe the relationship between energy use and economic
development shown in Figure 3.   (4 marks)

  (ii) Suggest TWO reasons for this relationship.   (6 marks)

(b) Show how political factors influence energy supply.   (6 marks)

(c) With reference to one or more located examples, examine how the
exploitation of energy resources brings social and economic opportunities. (9marks)

■ ■ ■

D-grade answer

(a) (i) There are quite a few countries who have a low GDP and a low total energy
consumption. Then there are some in the middle like Poland. The richer countries use more energy. Canada uses the most because it is a cold country
in winter.   1/4 marks

✏ This is a poor response as it does not describe the relationship. Rather it
describes in very basic terms three groups of countries and their levels of wealth
and energy use. It also drifts into explanation with the comment about Canada. A
simple statement describing what happens to energy consumption as GDP
increases is required with some reference to groups or individual countries.

(ii) In poor countries people cannot afford to pay for energy and so they don't use very much. Also countries like India are hot and so people don't need central heating but in Canada it is very cold and so they have to use a lot of energy. Australia is a hot country but people there use energy for cooling. Also people in richer countries have cars and these use lots of energy.   **2/6 marks**

  e This is a very simplistic answer written in an unconvincing manner placing it in level 1. There are hints of valid points such as the ability to pay for energy and the different energy demands of countries experiencing contrasting climates. However, the candidate has not grasped the basic relationship and the reasons for it, such as energy consumption in the manufacturing of consumer goods and the use of human and animal labour.

(b) There are lots of factors influencing energy supply and political factors are important. Some governments decide to concentrate on nuclear energy and build nuclear power stations. France has many nuclear power stations and it gets about 75% of its energy from this. In the UK we have been closing nuclear power stations but have not started building new ones yet.

If there is a war in the Middle East the price of oil goes up. This is because the Middle East supplies vast amounts of oil and if their countries are fighting then they will not be able to produce as much oil. In Iraq for example it is very difficult to produce oil because of the fighting. Some countries are going for renewable energy such as wind farms. These don't use fossil fuels and produce carbon dioxide but some people don't agree with them as they spoil the countryside. The government can build a wind farm even if local people don't want it.   **4/6 marks**

  e This answer offers three valid points related to political factors but needs to explain their respective influences in clearer language. It is important that the link between the factor and energy supply is explicit and not left for the examiner to deduce. That said, there is sufficient knowledge and understanding in the answer to lift it to the top of level 1.

(c) In China they have built the Three Gorges dam. This is on the Yangtze River which flows into the sea at Shanghai. One of the reasons the dam was built was to give hydroelectricity to China. China is developing very quickly and needs lots of energy for industries. The dam will provide about 4% of China's energy which doesn't sound very much but is a lot as it is such a large country. Many of these industries are in Shanghai and they produce goods for export. China needs to keep these industries growing so that they improve their economy. The dam also helps prevent the flood hazard along the Yangtze. When the dam was finished a huge lake built up behind it. It is over 600 kilometres long. This meant that many towns and farmland were flooded. However new towns were built with higher-standard houses and so this is a social opportunity for the people. They might also get jobs in the power station or to do with the increased trade along the river now that it is less dangerous for boats.   **6/9 marks**

ℓ It is good that there is a clearly identified example which is appropriate for this question. The links between the dam and social and economic opportunities are evident although not always clearly expressed. The structure of the answer would be improved by the use of paragraphs to separate the different points. Some of the terminology is rather inaccurate but overall the answer is placed in the middle of level 2.

ℓ **The total score for this four-part answer is 13/25, a D grade.**

■ ■ ■

A-grade answer

(a) (i) There is a strong positive relationship between GDP per capita and energy consumption per capita. The more wealthy countries such as France and Australia have higher energy consumption than the poorer ones such as Ghana and India. Canada is an anomaly as it uses so much energy.   **4/4 marks**

ℓ This a strong opening answer. The relationship is clearly described and then an appropriate amount of detail added to confirm the candidate's understanding.

(ii) As economies develop they need more energy. Energy is required to manufacture goods which are used to improve standards of living. For example it takes energy to build infrastructure such as roads and water systems. Energy is used to produce food which means that people are better fed. Travel involves using energy which allows goods to be transported which helps increase wealth.   **5/6 marks**

ℓ The explanation of the relationship constitutes a level 2 response as it focuses on that relationship and offers some relevant points. It is interesting that the candidate looks at the more developed perspective and therefore only implies something about the rest of the relationship. Some mention of why lower-income countries have low energy consumption would lift this to full marks.

(b) Energy supply is strongly influenced by political factors. In many countries the government is in charge of energy supply. Therefore decisions about what type of energy and where to locate power stations are political. Recently Sweden has decided to focus on nuclear energy rather than using fossil fuels. This is interesting as Sweden is a very 'green' country but the government has decided that they should not use energy that produces greenhouse gases. They are also investing more in renewables such as wind turbines.

Oil is one type of energy very influenced by political factors. OPEC is an organisation which controls much of the oil production. Its members try to get as high a price for their oil as possible. When the price of oil drops too low they will cut production to try to raise prices.   **6/6 marks**

> 🖉 This is an effective level 2 answer. It uses two relevant generic points with some helpful place detail in support. The candidate applies his/her knowledge and understanding appropriately to two different scales, national and international.

(c) Nigeria is an oil-rich country in West Africa. Nigeria is the fourth-largest exporter of oil in the world with nearly all of its production going for export. Natural gas is also exported in liquefied form. This means that Nigeria receives a vast amount of revenue from oil and gas. This can be used to help improve the country's economy and help pay for health and education. There are jobs in the oil industry for local people although the use of transnational companies means that much of the money does not stay in Nigeria.

In the UK North Sea oil and gas has developed since the 1970s. There have been many jobs created and places such as Aberdeen have really benefited. The TNC oil companies operating the drilling rigs need labour but so do all the industries who supply goods and services to them such as the helicopter operators and there is a multiplier effect in Scotland.

Also young people living in the area can find jobs and so they don't leave. This gives a more balanced community and helps the local economy.

In some countries the exploitation of energy resources can really help a remote region. In Alaska oil is obtained from the North Slope and this brings in money and jobs to a region which doesn't have many. The US government is keen to allow more oil exploitation as it then doesn't need to import as much from regions such as the Middle East.

Nearly always energy exploitation brings economic and social opportunities but it can also give disadvantages to people and cause pollution.   8/9 marks

> 🖉 The answer considers a range of relevant and quite detailed social and economic opportunities brought by the exploitation of energy resources. The answer is well structured into three paragraphs, each focused on a different example. There is some repetition of the basic point about economic opportunities but it is made at different scales, for example national, regional and for individuals. A couple of actual figures, such as the proportion of Nigeria's GDP resulting from energy, would secure maximum marks.

> 🖉 **The total score for this four-part answer is 23/25, a very secure A grade.**

The growth of tourism

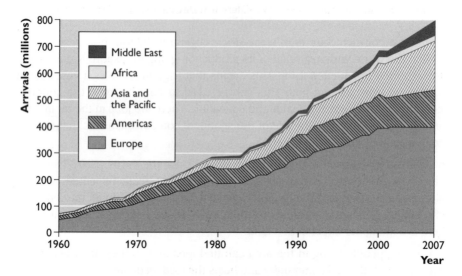

Figure 4 Change in international tourist arrivals in various regions of the world, 1960–2007

(a) (i) **Describe the changing pattern in tourist arrivals as shown in Figure 4.** (4 marks)

   (ii) **Suggest two reasons for this pattern.** (6 marks)

(b) **Show how tourism can play a significant part in the economic development of an area.** (6 marks)

(c) **With reference to one or more located examples, examine how an over-dependence on tourism can create problems for people.** (9 marks)

■ ■ ■

C-grade answer

(a) (i) It is clear that a huge increase in tourism has taken place. Europe has gone up from about 50 millions to about 400 millions. There is also an increase in the Americas to 500 millions and Asia and the Pacific is also up a lot. Not so many tourists go to Africa and the Middle East. 2/4 marks

   e This level 1 answer correctly identifies the overall increase in tourism in terms of tourist arrivals with the first sentence. The detail provided about arrivals to Europe is encouraging. However, from then on the answer deteriorates. The candidate is not able to interpret the graph correctly as the statement about the Americas indicates. It is important that you practice interpreting the full range of graph styles likely to appear in the exam paper so that elementary errors are avoided.

   (ii) More people have higher wages so can afford to travel abroad for holidays. Many Americans come to Europe for its history and culture and to visit relations. People can also afford to go abroad rather than have a seaside holiday.

There are also many more aircraft nowadays than in the 1960s. This means people can travel abroad easily. Aircraft can fly faster which means you can travel to places further away. 4/6 marks

e This answer is heading in the right direction with its choice of appropriate reasons but it does not use the points as well as it might, taking it to the top of level 1. The first reason, higher wages, should be developed into disposable income as it is this economic factor rather than simply high wages that allows people to travel internationally. The second reason, number of aircraft, is relevant. However, factors such as the introduction of wide-bodied aircraft which give economies of scale and the absence of tax on aircraft fuel are more significant. The candidate then offers a third reason, the speed of aircraft. It is important not to muddle your responses so that the examiner is clear which your chosen reasons are.

(b) Tourism can play a very significant part in the economic development of an area. In Thailand there has been a massive increase in tourism. In 2006 11.5 million people visited Thailand. Tourists stay in hotels which bring income to the country. There are many people employed by the hotels such as waiters and cleaners so their wages are paid by the tourists. Tourists pay taxes which helps the country. Tourists spend money when they are holiday and this gives the country more money. However, there is the problem of leakage which means tourist money does not stay in the country. 4/6 marks

e This is a level 1 answer as it includes some suitable points but does not develop them to their full potential. There are too many vague statements such as 'Tourists pay taxes which helps the country', which is true but the candidate does not relate this explicitly to the economic development of Thailand. The last sentence indicates the candidate has drifted away from the question as the issue of leakage is irrelevant here. Also, the first sentence is not needed in the context of these short-answer questions. It takes valuable time away from writing about the specific factors.

(c) In Gambia there is a very big tourist industry. It employs lots of people both officially and unofficially. Most of the tourism is located along the coast at places such as Banjul. Gambia is a very poor country with a low GDP, only $250. The government tries to use tourism as a way of developing the country. The problem they have is that most of the money from tourism goes out of the country called leakage. This is because most of the hotels are owned by foreign firms and they take the profits back to their country. There are wages and taxes paid to Gambia but not as much as is lost by leakage.

There are two seasons in Gambia, wet and dry and tourists only really want to go there during the dry season, November to April. This is when most visitors come from Europe for the winter sun holidays. During the other months fewer tourists come and so there is unemployment.

4

question

The jobs in tourism are not always well paid. If you are a waiter or cleaner then your wages are not very high. Most of the high-paid jobs are given to foreigners.

In some parts of the coast there has been destruction of ecosystems due to building hotels. The beaches can be ruined with too much pollution. The Gambian government is trying to do something about this by developing ecotourism for wildlife. 5/9 marks

e The extended answer sub-part will often allow the use of one or more 'located examples'. You are not at a disadvantage by using either of those you have learned, but if you focus on 'one' then give plenty of detail. This answer chooses to concentrate on Gambia, a suitable example, but overall the level of detail is not convincing. The point about leakage is valid and is soundly explained but does not explicitly link with over-dependence on tourism. The mention of seasonal unemployment is good and moves the answer forward. The last point about the environment has potential but is vague in its expression. The answer ends up as a level 2 response.

e **The total score for this four-part answer is 15/25, just a C grade.**

■ ■ ■

B-grade answer

(a) (i) International tourist arrivals have increased from under 100 million in 1960 to over 700 million in 2007. The most significant increase has been since 1980. All areas have increased, Europe is the largest destination but Asia and the Pacific is increasing rapidly. 4/4 marks

e This is a strong level 2 response to open with. An accurate summative sentence is followed by detail confirming that the candidate understands the resource and the pattern.

(ii) Since 1960 there have been several reasons influencing tourism. The first is disposable income. People now have more money to spend on holidays with some people able to take more than one holiday a year. People also have paid holidays which allows them to travel abroad.

The second reason is the development of the travel industry. Package holidays are now available from operators such as Thomas Cook. This makes it easy for people to travel abroad as all the arrangements are made by one company. Also people can now use the internet to book holidays. Airlines and hotels advertise on-line and people can arrange their own holidays. 6/6 marks

e This top of level 2 answer is well constructed and shows secure understanding of two relevant reasons. Both are explained so as to answer the question directly. The use of paragraphs aids the answer as it makes clear the separate reasons.

(b) The economic development of an area can be helped by the development of tourism. Tourists spend money when they are on holiday and this brings money into an area. This money means that wages are paid and so local people have money to spend. When they spend money other people have jobs. This is called the multiplier effect. *2/6 marks*

> After two strong answers this is a very disappointing level 1 response. The candidate offers only vague comments, which although relevant, are not developed into a convincing argument. The lack of any place detail detracts from the answer; it is often helpful to set an answer to a question like this in a real-world context.

(c) Tourism can bring both advantages and disadvantages to people. In the Canary Islands, a group of Spanish islands off the north-west coast of Africa, there are both advantages and disadvantages of tourism.

The biggest island is Tenerife and tourism is the most important sector of the economy. About 80% of the island's economy depends on tourism. This is not a problem when tourism is booming but when fewer people visit Tenerife then there is not much else in the economy. People are unemployed and the government collects less taxes to pay for schools, health care and roads.

Although there are plenty of jobs in tourism they are usually not well paid. Waiters in restaurants and hotel staff such as cleaners are not well paid and can easily be made unemployed if the number of visitors declines.

Many people, especially the young, move to the main towns such as Los Cristianos. They leave rural areas so that there is a decline in agriculture. In some parts of the island, farmland has been abandoned and small villages struggle to survive. This leaves the older people with not enough shops and services.

Finally there is a problem with education. If young people think that they are going to get a job in tourism they may not bother too much about their education. This is because most of the jobs do not require highly qualified staff.

Although there are many advantages to tourism on Tenerife there are also problems as it is over-dependent on tourism. *8/9 marks*

> This is an encouraging level 3 answer that offers four valid explanations that are focused on the question. The answer is clearly set in a suitable place context and gives some convincing data about Tenerife. Although the answer is well structured via its use of paragraphs, in places the quality of expression is not as effective as it might be. For example, there could be greater precision given by the use of geographical terminology such as 'threshold' in the paragraph about rural decline. But overall this is a good answer.

> **The total score for this four-part answer is 20/25, a top B grade. If the third sub-part had been more detailed this would be a clear grade-A response.**

uestion 5

Section B: extended writing questions

Managing urban change

With reference to one or more located examples, examine the problems in managing changing demand for services such as health, education and public transport.

(25 marks)

■ ■ ■

C-grade answer

In Nairobi there is rapid population growth due to natural increase and migration from the countryside. Nairobi is the capital of Kenya and is where many local people want to live. They move to the city as they hope to find jobs which will pay them more than they were earning in the countryside where they were sub-sistence farmers. One of the problems they find when they arrive in Nairobi is housing as they can't afford to pay very much for it. This means they end up living in slum areas like Cubera which is the largest slum area in Africa. The houses are built to a low standard as they can't afford to use proper materials so they use things such as plastic sheeting, metal pieces on the roof and scraps of wood. Also many use mud for the walls as it is free.

The parents hope that the children will go to school and receive a good educa-tion so that they can have better lives. There are some schools but not enough so many of the children do not go to school but find part-time jobs selling things on the streets. The Kenyan government doesn't have enough money to build more schools but there are non-government organisations (NGOs) who are charities such as Christian Aid and Oxfam helping in the country. They provide some schools but it still isn't enough as the population is growing so fast.

There are also problems in providing health care as the government lacks money. Conditions in the slums are terrible with no proper drainage. This means that dis-ease such as cholera can spread quickly as it is so overcrowded. TB is also very common as the houses are so crowded together that the air doesn't move as much and so the risk goes up. There is no waste collection in Cubera as the city council says the slum area doesn't exist! This means that waste piles up all over the slum causing more disease. This attracts rats and flies which spread more disease.

The demand for services keeps on going up in Nairobi but the government finds it very difficult to meet it. NGOs try to help but they can only make a small differ-ence but there are plans to improve things like sewage and housing.

e This answer contains some appropriate knowledge and understanding but opens with a largely irrelevant paragraph about housing. The first couple of sentences are fine as they establish the 'changing demand' via a growing population. The details about building materials are irrelevant here. It is also important to spell place names correctly, Kibera not Cubera! The remainder of the answer is suitable as it looks at the services of education, health care and sanitation. The answer reaches just into level 2 for knowledge and understanding.

There is some analysis of the problems, the points about rapid growth and the council denying the formal existence of the slum areas for example. But the lack of a sharp focus on the wording of the question mean that level 2 only is achieved for analysis and application.

Finally the answer is well structured with the use of paragraphs but it is not sufficiently secure in its spelling and grammar so level 2 is achieved again here.

e **The answer scores 8 marks for knowledge and understanding, 4 marks for analysis and application and 4 marks for skills and communication, making 16 in total, a C grade.**

A-grade answer

Although urban areas are very different in MEDCs and LEDCs, they both have problems with providing services such as health, education and public transport. I will use the examples of Birmingham and Mumbai in my answer.

Birmingham's population grew until the 1970s when depopulation took place. This happened especially in the inner city such as Lozells and Small Heath. Then the terraced housing that was left was occupied by large numbers of migrants, mainly from Asia or the Caribbean. They were attracted to the inner city as it was cheap and relatively close to employment in and around the CBD. There have been problems in providing education to these communities. One of the difficulties has been language as not enough teachers speak the various languages spoken by the children. Also many of the parents do not have very good English and so it is difficult for the schools to involve the parents. Birmingham City council has run language courses for both teachers and parents which have helped. Many of the schools have had building projects to improve their facilities such as making safer playgrounds and upgraded classrooms.

Health care is a problem in inner-city areas as few doctors and dentists want to work there. New health centres have been built to provide medical services to the local people. Staff have been trained in some of the languages and leaflets are printed in many different languages. It is also important that Muslim women see female doctors and the health service has done this.

The Heartlands redevelopment programme has been offering grants for housing improvements such as damp-proofing the older terraced houses. There has also been some new housing built for example in Small Heath which have been built as traditional houses not tower blocks.

In Mumbai the problems all centre on the rapidly growing population. There is insufficient housing and many of the poorer people cannot afford to pay high rents. So they migrate to the slum areas called bustees in India such as Daravi. Here people live at incredibly high densities, over 100 000 per km² which causes health problems. Most of the people in these areas cannot afford medical care and there are not enough doctors. As well, very few doctors want to work in the slum areas.

It is interesting that although different, both Birmingham and Mumbai have problems managing service provision especially to their poorer inhabitants. It seems that poverty is an important factor when dealing with inequality.

e The answer addresses the question very effectively, using a combination of located examples, one MEDC and one LEDC. This contrast in type of changing demand indicates a thoughtful approach although there is a weakness regarding the place detail for Mumbai. Overall, factual content and understanding and cause and effect are convincing enough for level 3.

There is clear analysis of the management strategies being employed to deal with a variety of problems and the response is well structured with accurate use of spelling, grammar and geographical terminology.

e **The answer scores 12 marks for knowledge and understanding, 5 marks for analysis and 7 marks for skills and communication. This gives the impressive total of 24/25, a very strong A grade.**

Managing rural change

With reference to one or more located examples, explain how rural areas can be managed to ensure sustainability. (25 marks)

■ ■ ■

D-grade answer

In the lake District there are thousands of visitors every year. They started going to the lake District in the 19th century when William Wordsworth wrote the first guide to the area. When the railways arrived many people were able to travel there and towns such as Windermere grew rapidly. Then the car was invented and even more people visited the lake District and people can travel all over the region. Today about 8 million visitors come to the lake District. All these people cause problems to the area although they bring in much income for the area. When lots of visitors go walking the footpaths are eroded. All there boots wear away the top soil meaning that water and wind erosion can take place. This causes gullying and the footpath becomes wider. One of the modern problems is mountain biking along footpaths as these also cause erosion with the tyres.

The other problem in the lake District is the traffic. This is because the roads are narrow and there is to much traffic. We saw this when we visited Windermere on a field trip in October. Most visitors come by car and the roads cannot cope. There is also the issue of car parking as there is not enough. The National Park wants to raise the parking fees and introduce a shuttle bus. This will carry visitors from a car park to a honey-pot site such as lake Windermere. The farmers in the lake District have sheep. They were paid by the EU to have as many sheep as possible which caused soil erosion through the trampling by the sheep. However, this is not happening as much as the way the farmers are paid has changed and not so many sheep are kept.

There is some sustainable parts to the lake District. For example the lakes are zoned for the different people using the lake. There is a speed limit now so that boats do not cause problems for fishermen and sailors.

e This is on the short side for an answer in this section of extended writing worth 25 marks. The candidate nowhere indicates that he/she knows what is meant by sustainability neither is there any sustained indication that management practices are securely known and understood. The choice of the Lake District is appropriate but there is a lack of detail. The answer reads as if it would be better in the 'Growth of tourism' option. It is quite in order to use the same case study in more than one option, but the answer must be written so that it clearly answers the particular question set. This candidate does not manipulate the material to suit a question in the 'Rural' option. Level 1 for knowledge and understanding is all that can be awarded. There is some analysis and application lifting it to the bottom of level 2 and the same is true for skills and communication. There is some structure with the use of paragraphs and some of the grammar is correct. In places spelling and use of terminology is inaccurate.

✏ **The answer scores 5 marks for knowledge and understanding, 3 marks for analysis and application and 4 marks for skills and communication. This gives a total of 12/25, a bottom D grade.**

A-grade answer

Sustainable development is usually taken as development which meets the needs of the present without compromising the needs of people in the future to meet their needs.

Many rural areas in MEDCs have suffered from various problems which management has been trying to solve. In northern Sweden the population density is only 2 persons per km² and most of the area is remote and inaccessible. Most of the employment is in primary industries such as farming, forestry and mining which is a problem as it is a narrow range of economic activity. There has been much migration of young people to towns such as Sundsvall and even right out of the region. This has left an aging population in the region. There is, however much energy from the 22 HEP plants in the area.

Although forestry may not be sustainable, Sweden is managing its forests sustainably. For a long time they have been planting more than was being cut down and now other uses of the forests are being encouraged such as recreation. Sweden relies a lot on its forests for income and they are managing them sustainably. The authorities are trying to encourage tourism into Västernorrland as this brings in income and gives employment which may help keep some of the younger people. A World Heritage Site has been made and there is a national park and many nature reserves. All this is trying to make sure that tourism will be sustainable for the future.

In East Anglia farming is now being run more sustainably. In the past farmers received large grants and subsidies from the CAP of the EU. This meant that they switched into large-scale arable farming. This led to hedgerow removal, draining of wetlands and soil erosion. Changes have been taking place to the CAP so that environmental sustainability is more important. Set-aside is a scheme that reduces the area of a farm used to grow crops. Instead the land is either unused or grass is grown. No agrochemicals are allowed so this all helps the natural ecosystems recover. This can be seen in the numbers of birds such as lapwings and partridges.

The water quality has also been improved to help sustainability. There used to be a lot of leaching of nitrates caused by chemical fertilisers. Nitrates in the water can cause algal blooms and in drinking water have been linked with some diseases. Now farmers are not allowed to put too much fertiliser on their fields nor can they apply it before March 1st. This is because nitrates are very soluble and the winter rains wash it off the fields. All this is done to make farming more sustainable.

e The candidate has chosen to use two contrasting case studies which give place-specific material, and avoids repetition and ineffective generalisation. There is a good depth of knowledge in the case studies although the East Anglian example would benefit from a little more place detail. For this answer, knowledge and understanding level 3 is awarded. Throughout the answer the candidate keeps a sharp focus on the issue of management for sustainability and this degree of relevance results in level 3 for analysis. The answer also achieves level 3 as it is well written with an appropriate structure (although the conclusion is too brief), and accurate spelling and use of geographical terminology.

e **The answer scores 12 marks for knowledge and understanding, 5 marks for analysis and 6 marks for skills and communication. This results in a total of 23/25, a strong A grade.**

The energy issue

With reference to one or more located examples, examine problems the exploitation of energy resources brings for the environment. (25 marks)

C-grade answer

There is often a lot of pollution caused by the exploitation of energy resources.

When fossil fuels such as coal and oil are burnt gases such as carbon dioxide are given off. This causes damage to the atmosphere as these gases help cause global warming. When this happens temperatures rise and the ice sheets melt. Sea level rises and causes flooding along coastal areas. This means that more money needs to be spent on coastal protection using hard and soft engineering and managed retreat. This has happened along the Wash in East Anglia.

The Chinese have built the Three Gorges Dam on the Yangtze River. This is a massive project that produces a great deal of renewable energy as HEP. About 4% of China's electricity comes from this one project. Although it is renewable energy there are still problems for the environment. The lake behind the dam has flooded vast areas, about 1000 km² along the river which has caused difficulties. The water in the lake is fairly still and so pollution levels rise as the water does not carry them away. When the lake was formed there were lots of towns and industries flooded such as Wushan. These have caused pollution as waste materials have now leaked into the lake. In some places coal mines have been flooded and this causes more pollution as waste left behind by mining has been washed away. This has meant that rare species such as the Yangtze dolphin have been made extinct. Local fishermen do not catch as much as before the dam was built. Algal blooms can result from too much pollution as the oxygen in the water is used up. The whole ecosystem can be damaged. The farmers who used to live along the river have had to move as the lake built up. Their fields are now under water and so they have had to move. The Chinese government has resettled them but some farmers say that the new farms are not as fertile as the ones they had before. As the river does not flood anymore they will have to use artificial fertiliser as the silt the river carried is trapped behind the dam.

The answer shows some knowledge and understanding of the topic. Its main focus, the Three Gorges Dam, is an appropriate located example and in this section the candidate tries to establish cause and effect between the dam's construction and formation of its vast lake and environmental problems. However, much of this material is not as detailed as it could be, for example more facts and figures could be included. It is not always clear if the candidate is referring to the area upstream of the dam or below it, an important distinction that could help lift an answer into level 3.

The opening paragraph is not without merit but the cause and effect mentioned is not wholly convincing: thermal expansion of water is more significant than ice melt. The answer also needs more detail within the example if it is to be effective. Overall this answer reaches level 2 for knowledge and understanding.

There is some analysis of how the exploitation of the hydroelectric potential of the Yangtze impacts on the environment but this is too tentative and more secure links need to be explained. Again this results in a level 2 mark.

For skills and communication the answer reaches level 2. There is a degree of structure but more should be applied to the material on the Three Gorges Dam, such as paragraphs. There is an opening sentence but no conclusion.

e **The answer scores 8 marks for knowledge and understanding, 4 marks for analysis and application and 4 marks for skills and communication. This gives a total of 16/25, a sound C grade.**

A-grade answer

The exploitation of energy resources creates many opportunities but it can also bring problems. These problems can affect the environment quite badly although in some cases management can reduce their impact.

One area where energy exploitation has definitely had a negative influence on the environment is in the Niger Delta. The Delta is located in Nigeria where the river has deposited millions of tonnes of alluvium. Under this sediment are vast reserves of oil and natural gas which are being extracted. In 2008 about 135 million tonnes of oil were extracted and natural gas production was about 34 billion cubic metres. Most of the exploration and production involves TNCs such as Shell.

One major environmental problem is oil spills. When oil is drilled some oil is almost inevitably spilled but in the Niger Delta there is very little environmental protection. The system of pipes which connect the oil wells with where the oil is refined and stored such as Port Harcourt is the source of most of the spills. Pipelines quickly corrode in the very humid tropical climate of the Delta and because there are so many pipelines, maintenance is poor. If the pipeline is near one of the waterways in the Delta the oil can quickly spread over a large area. This destroys the aquatic ecosystem and local fishermen have their catches reduced. In particular damage is done to the mangrove ecosystem which is an important habitat in the delta. It also helps protect the coastline against coastal erosion as the trees absorb wave energy. Oil spills have also damaged farmland meaning that subsistence farmers have their land ruined.

Another environmental issue is the flaring of gas. When oil is drilled there is usually gas with the oil. This is burnt off as a waste product. When this happens it produces air pollution such as acid rain and creates fogs. Local people suffer more from respiratory diseases and cancer due to air pollution.

question

The issue of oil spills is not just because the oil companies have not maintained the pipelines properly. There is the illegal tapping of oil from the pipelines by local people. When this happens spills are more likely and the risk of explosion very high. If a pipeline explodes people are often killed but also huge amounts of oil spill onto the land until the oil company can get there to repair it.

The drilling for oil and gas in Nigeria has brought much wealth to the country but there are also major environmental problems. The oil companies are responsible to some extent but then so are the local people.

e This is a strong answer focused on the question. It uses one example which is fine, as long as plenty of place detail is given. The knowledge and understanding contained in the answer is detailed with the cause and effect of oil and gas exploitation in the Delta discussed effectively. A number of environmental problems are included and are sensibly analysed. The candidate uses a sensible structure with a brief introduction, three paragraphs of place-specific material and then a simple conclusion. All three elements of the assessment reach level 3.

e **The answer scores 12 marks for knowledge and understanding, 5 marks for analysis and application and 6 marks for skills and communication. The total mark is 23/25, a strong grade A.**

Question

The growth of tourism

With reference to one or more located examples, explain how sustainable tourism balances socioeconomic and environmental needs. (25 marks)

B-grade answer

Tourism in LEDCs may be seen as a great way to quickly increase the amount of foreign exchange to a country. However the environmental impacts of tourism may not always be considered as the country or island is overwhelmed by tourism. However some countries do decide to bring in development plans for ecological safeguards to try and reduce these impacts rather than opting for total tourism.

Mass tourism in an LEDC can lead to the destruction of natural habitats, flora and fauna in order to create space for building developments. A good example of this is the destruction of mangrove swamps in the carribean in order to drain them and build on them. In this situation there needs to be consideration of the wealth they could and do provide.

Another impact on the environment caused by mass tourism is the obvious foot-path erosion due to huge numbers of tourists visiting an area and wanting to explore by foot. This occurred for example in the Nepalese foothills. Along the Himalayas various ecotourism programmes have been brought in to try to reduce the effect of tourists on the environment such as the Annapurna project. For example they now have solar showers which use renewable energy instead of burning wood. There is also a fee charged for anyone who visits Nepal and the money from this goes to conservation projects. They have set up tree planting programmes to try to stop soil erosion from the heavy monsoon rains. The trees intercept the rain and their roots hold the soil together and stop it being washed away. There is also education of the visitors about the environment and this helps with conservation.

In Costa Rica there are eco-lodges for people to stay in when they visit the rainforest. At Selva verde you can stay in buildings built from local materials which are on stilts. They are designed to blend in with the rainforest. There are long elevated walkways that allow you to explore the rainforest without very much disturbance of the ecosystem. You can also stay with local families and learn about their culture and how they live and work in the forest.

There is more being done to make tourism sustainable especially in LEDCs.

e The answer refers to two examples, both from LEDCs, and offers some details regarding the management strategies followed to establish sustainable tourism. The content has some relevance and in places directly links tourism with the strategy. However, there is scope for more place detail and at times the answer wanders away from a sharp focus on the question. The concept of sustainability is implied rather than dealt with explicitly; it is a good idea to offer a brief definition

questions & answers

of terms/concepts particularly when they are directly mentioned in the question. The answer is quite well written, using paragraphs to give a structure. There are some grammatical and spelling errors and the conclusion is limited.

@ **The answer scores 10 marks for knowledge and understanding, 3 marks for analysis and application and 5 marks for skills and communication. This gives a total of 18/25, a B grade.**

A-grade answer

Tourism has developed into one of the world's largest industries. Globally it has a value of nearly US$700 billion. It also relies on a range of resources, natural, cultural and heritage. If these are exploited so that they are damaged they will not provide sustainable tourism. This type of tourism aims to provide for the needs of today but without exhausting the resources so that there is no tourism in the future.

It is therefore very important that the tourism developments are sustainable. Governments are becoming more directly active in helping to make tourism sustainable. In Bhutan, a limit of just under 3000 tourists is allowed entry each year. This makes sure that a carrying capacity is set and stuck to ensuring sustainability as less pressure is put on the environment. There is also a surcharge paid by every tourist of £75 in order to generate revenue to reinvest back\ in the country.

In terms of the environment, sustainability is very important. Instead of mass tourism such as in the Spanish Costas, alternative or ecotourism is used. Ecuador runs the Galapagos Islands in the Pacific Ocean about 1000 km from the coast of South America. The islands are extinct volcanoes and because of their geographical isolation the ecosystems are very fragile. However, Ecuador is a relatively poor country and needs the revenue that tourism brings. They did not want to stop tourism altogether so came up with a plan to keep tourism sustainable. Most of the islands have been made into a national park. This helps give the rare species protection such as the famous giant tortoises, marine iguanas and blue feet boobies, a rare type of sea bird. There is also an entrance fee to visit the national park and this money helps pay for conservation measures. The government may set the fee much higher to raise more money. They are also concerned about oil spills from cruise ships and fishing boats but it is difficult to stop this happening.

Sustainability is also found in MEDCs. In Arches National Park, Utah, USA nearly 1 million visitors come to see the natural stone arches and desert scenery. Most of the people come between May and September when there is severe road congestion and overcrowding at the most popular sites such as Double Arch. Cars parked off the road damage the fragile desert soil and there is footpath erosion along some of the trails. There are plans to have reserved parking at some of the key attractions to limit numbers and more patrols to stop illegal parking.

In Dartmoor National Park there are also concerns about visitor numbers and the different types of activities. The park authority has a zoning policy in place that encourages certain activities in some sites and discourages others. Visitor centres have been built in some locations that are under the most pressure such as at Hay Tor. Here more footpaths have been created so that not everyone follows the same path which has been leading to erosion. In some parts of the moor, people are discouraged from going there during the breeding season so that birds are not disturbed. There is also a Code of Conduct for climbers so that the tors and the area around them are not eroded.

It is important that tourism is made sustainable as so many people rely on it for their employment. In some places it may be that visitors have to be restricted in their numbers or where they can go so that there is still a tourist industry in the future.

e This is a wide-ranging discussion that keeps its focus on the question. Its breadth rather than depth approach is successful and offers some appropriate examples of sustainability issues and management responses. The case studies complement each other with the inclusion of MEDC examples being particularly encouraging. Too often certain topics are only seen in one or other development contexts. It is well written with an effective structure that includes both an introduction and conclusion. The answer is clearly level 3 for all three assessment objectives.

e **It scores 13 marks for knowledge and understanding, 5 marks for analysis and application and 7 marks for skills and communication. This gives a splendid 25/25 marks, a highly convincing A grade.**

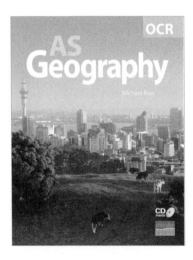
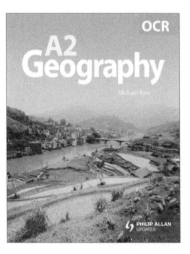

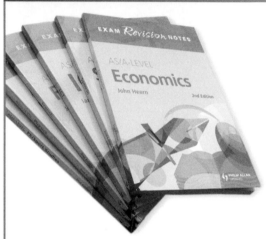

PHILIP ALLAN
UPDATES

INTENSIVE
REVISION
WEEKENDS

Intensive Revision Weekends

Biology

Business Studies (AQA)

Chemistry

Economics

French

German

Law (AQA or OCR)

Mathematics

Physics (Edexcel)

Politics

Psychology (AQA (A))

Religious Studies

Sociology

Spanish

- Intensive 2-day revision courses with expert tutors, including senior examiners
- Guaranteed to improve exam technique and build confidence
- Develops AS and A2 key skills
- Tests understanding of core topics
- Invaluable summary notes provided
- Ideal Central London location

Contact us today for further information: tel: 01706 831002 fax: 01706 830011
e-mail: sales@philipallanupdates.co.uk

Philip Allan Updates
Suite 16, Hardmans Business Centre, Rossendale, Lancashire BB4 6HH